Advances in Spatial Science

Springer
Berlin
Heidelberg
New York
Barcelona
Hong Kong
London
Milan
Paris
Tokyo

Titles in the Series

C. S. Bertuglia, M. M. Fischer and *G. Preto* (Eds.)
Technological Change, Economic Development and Space
XVI, 354 pages. 1995. ISBN 3-540-59288-1 (out of print)

H. Coccossis and *P. Nijkamp* (Eds.)
Overcoming Isolation
VIII, 272 pages. 1995. ISBN 3-540-59423-X

L. Anselin and *R. J. G. M. Florax* (Eds.)
New Directions in Spatial Econometrics
XIX, 420 pages. 1995. ISBN 3-540-60020-5 (out of print)

H. Eskelinen and *F. Snickars* (Eds.)
Competitive European Peripheries
VIII, 271 pages. 1995. ISBN 3-540-60211-9

J. C. J. M. van den Bergh, P. Nijkamp and *P. Rietveld* (Eds.)
Recent Advances in Spatial Equilibrium Modelling
VIII, 392 pages. 1996. ISBN 3-540-60708-0

P. Nijkamp, G. Pepping and *D. Banister*
Telematics and Transport Behaviour
XII, 227 pages. 1996. ISBN 3-540-60919-9

D. F. Batten and *C. Karlsson* (Eds.)
Infrastructure and the Complexity of Economic Development
VIII, 298 pages. 1996. ISBN 3-540-61333-1

T. Puu
Mathematical Location and Land Use Theory
IX, 294 pages. 1997. ISBN 3-540-61819-8

Y. Leung
Intelligent Spatial Decision Support Systems
XV, 470 pages. 1997. ISBN 3-540-62518-6

C. S. Bertuglia, S. Lombardo and *P. Nijkamp* (Eds.)
Innovative Behaviour in Space and Time
X, 437 pages. 1997. ISBN 3-540-62542-9

A. Nagurney and *S. Siokos*
Financial Networks
XVI, 492 pages. 1997. ISBN 3-540-63116-X

M. M. Fischer and *A. Getis* (Eds.)
Recent Developments in Spatial Analysis
X, 434 pages. 1997. ISBN 3-540-63180-1

R. H. M. Emmerink
Information and Pricing in Road Transportation
XVI, 294 pages. 1998. ISBN 3-540-64088-6

F. Rietveld and *F. Bruinsma*
Is Transport Infrastructure Effective?
XIV, 384 pages. 1998. ISBN 3-540-64542-X

P. McCann
The Economics of Industrial Location
XII, 228 pages. 1998. ISBN 3-540-64586-1

L. Lundqvist, L.-G. Mattsson and *T. J. Kim* (Eds.)
Network Infrastructure and the Urban Environment
IX, 414 pages. 1998. ISBN 3-540-64585-3

R. Capello, P. Nijkamp and *G. Pepping*
Sustainable Cities and Energy Policies
XI, 282 pages. 1999. ISBN 3-540-64805-4

M. M. Fischer and *P. Nijkamp* (Eds.)
Spatial Dynamics of European Integration
XII, 367 pages. 1999. ISBN 3-540-65817-3

M. M. Fischer, L. Suarez-Villa and *M. Steiner* (Eds.)
Innovation, Networks and Localities
XI, 336 pages. 1999. ISBN 3-540-65853-X

J. Stillwell, S. Geertman and *S. Openshaw* (Eds.)
Geographical Information and Planning
X, 454 pages. 1999. ISBN 3-540-65902-1

G. J. D. Hewings, M. Sonis, M. Madden and *Y. Kimura* (Eds.)
Understanding and Interpreting Economic Structure
X, 365 pages. 1999. ISBN 3-540-66045-3

A. Reggiani (Ed.)
Spatial Economic Science
XII, 457 pages. 2000. ISBN 3-540-67493-4

D. G. Janelle and *D. C. Hodge* (Eds.)
Information, Place, and Cyberspace
XII, 381 pages. 2000. ISBN 3-540-67492-6

P. W. J. Batey and *P. Friedrich* (Eds.)
Regional Competition
VIII, 290 pages. 2000. ISBN 3-540-67548-5

B. Johansson, Ch. Karlsson and *R. R. Stough* (Eds.)
Theories of Endogenous Regional Growth
IX, 428 pages. 2001. ISBN 3-540-67988-X

G. Clarke and *M. Madden* (Eds.)
Regional Science in Business
VIII, 363 pages. 2001. ISBN 3-540-41780-X

M. M. Fischer and *Y. Leung* (Eds.)
GeoComputational Modelling
XII, 279 pages. 2001. ISBN 3-540-41968-3

M. M. Fischer and *J. Fröhlich* (Eds.)
Knowledge, Complexity and Innovation Systems
XII, 477 pages. 2001. ISBN 3-540-41969-1

M. M. Fischer, J. Revilla Dietz and *F. Snickars*
Metropolitan Innovation Systems
VIII, 270 pages. 2001. ISBN 3-540-41967-5

L. Lundqvist · L.-G. Mattson (Eds.)

National Transport Models

Recent Developments and Prospects

Erratum

Due to an unfortunate error,
page 193 is missing in this book.
The text is printed on the reverse side
of this sheet.

Springer

Author Index

Lars Lundqvist · Lars-Göran Mattsson
Editors

National Transport Models

Recent Developments and Prospects

With 35 Figures
and 23 Tables

Springer

Professor Lars Lundqvist
Professor Lars-Göran Mattsson
Royal Institute of Technology
Department of Infrastructure and Planning
Division of Transport and Location Analysis
SE-100 44 Stockholm
Sweden

ISBN 3-540-42426-1 Springer-Verlag Berlin Heidelberg New York

Library of Congress Cataloging-in-Publication Data applied for
Die Deutsche Bibliothek – CIP-Einheitsaufnahme
National Transport Models: Recent Developments and Prospects / Lars Lundqvist; Lars-Göran Mattsson (ed.). – Berlin; Heidelberg; New York; Barcelona; Hong Kong; London; Milan; Paris; Tokyo: Springer, 2001
(Advances in Spatial Science)
ISBN 3-540-42426-1

Springer-Verlag Berlin Heidelberg New York
a member of BertelsmannSpringer Science+Business Media GmbH

http://www.springer.de

Printed in Germany

Hardcover-Design: Erich Kirchner, Heidelberg

SPIN 10848052 42/2202-5 4 3 2 1 0 – Printed on acid-free paper

Preface

National and international socio-economic integration processes have been important driving-forces behind the increase of interregional and international mobility of persons and commodities. This has led to new demands for analysis of long-distance transport. At the same time, the negative impacts of transport in terms of e.g. congestion and pollution and the competition for limited funding of new infrastructure have highlighted the need for more policy sensitive tools for transport demand forecasts and transport management.

This volume presents the state-of-the-art of national and European transport models at the turn of the century by presenting a sample of the most advanced national and European-wide models. These approaches represent wide differences in the scope and mode of modelling. A comparative analysis of these differences and some perspectives on future developments are provided.

The papers presented in this volume were first discussed at *The Second Seminar on National Transport Models* held in December 1998 at Långholmen Conference Centre, Stockholm. The conference was organised by PTRC Education & Research Services. A first seminar had earlier been arranged in the Netherlands in 1997.

The editing and comparative analysis have been supported by The Swedish Transport and Communications Research Board and by VINNOVA, The Swedish Agency for Innovation Systems. The final editing has been done most efficiently by Ingrid Johansson. The editorial process has been co-ordinated by Lars Lundqvist (from August 2000 with Master's Program in Environmental Sciences, University of Tsukuba, Japan) and Lars-Göran Mattsson, Division of Transport and Location Analysis, Royal Institute of Technology, Stockholm.

May, 2001

Lars Lundqvist
Tsukuba

Lars-Göran Mattsson
Stockholm

Contents

1 National Transport Models: Introduction and Comparative Analysis

Lars Lundqvist
Lars-Göran Mattsson
Department of Infrastructure and Planning
Division of Transport and Location Analysis
Royal Institute of Technology
SE-100 44 Stockholm, Sweden

1.1 Background

There are many reasons why national and European transport models have attracted increased attention during recent decades. The broadening of national transport policy from strategic infrastructure investments to infrastructure management with regard to efficiency, environmental, safety and regional equity objectives has lead to a need for advanced and more policy sensitive tools of analysis. The increase of interregional and international mobility requires forecasting tools that go beyond the urban or regional level. The competition for national infrastructure investments among regions and for Trans-European investments among nations has to be resolved by decisions and decision support systems at the appropriate spatial level. Environmental impacts transcend regional and national boundaries and transport policies affecting these environmental impacts involve all spatial levels.

It is the purpose of this chapter to discuss the state-of-the-art of national transport models in Europe. A number of national and European-wide models are compared. These models are presented in detail in separate chapters in this book.

The organisation of the chapter is as follows. The models are briefly introduced in Section 1.2. Section 1.3 presents a comparative analysis of the models with respect to five different aspects:

- model structure and theoretical approaches (Section 1.3.1);
- timetable based public transport assignment (Section 1.3.2);
- data and model estimation (Section 1.3.3);
- validation (Section 1.3.4); and
- areas of application (Section 1.3.5).

Finally the conclusions of the analysis and some ideas for future development are discussed in Section 1.4.

1.2 A Menu of Models

In many European countries national transport models are now available (e.g. in Sweden, Norway, Denmark, Germany, the Netherlands, the U.K., Italy and Hungary) (Gunn, 2001). Models are also developed to be applicable on a European scale. In some cases the development is quite new (as for the national models for Italy and Denmark). In Sweden the old system of models has recently been replaced by a new much more comprehensive system, SAMPERS. This system is now being implemented and applied by the planning authorities in Sweden. In the Netherlands the state-of-the-art national model system from the mid 1980's is now being thoroughly updated (Daly, 2001). The different national models have much in common – often using disaggregate nested (tree) logit structures and requiring similar kind of data. Their development has been stimulated by the exchange of ideas, knowledge, results and software among a group of modelling experts, consultants and clients that have met regularly and also formed specially dedicated fora for discussion.

The Dutch National Model System has served as a kind of 'prototype' disaggregate model in Europe – most clearly for the Norwegian national model (HCG and TØI, 1990). It was originally designed to be a tool for strategic appraisal of new road and rail links but its scope of application has gradually widened also to include e.g. environmental and IT issues. This kind of policy changes together with market changes, changes in attitudes, the base matrices becoming old, hardware and software technology advances, imply that the system now needs to be changed (Daly, 2001). Any development of an advanced model system such as this system is a large investment in money, time, effort and knowledge. The question is whether extra benefit can be derived from the investment by updating the existing system rather than developing a completely new system, and also whether a flexibility that would facilitate this even could be planned for when designing the model structure.

National travel demand models have been in use in Sweden since the beginning of 1980s. These models cover the usual kind of travel decisions but have not been integrated into one single comprehensive system. For several reasons Swedish planning authorities decided to order a new comprehensive and user-friendly system. The data, on which these earlier models were based, were getting old, 'new' modes like high-speed trains were not well represented in the earlier models, and some parts of the travel demand, e.g. long distance train trips, did not seem to be modelled in a reasonable way. In addition the scope and context of application have changed from the traditional one of providing demand forecasts for cost-benefit analysis of infrastructure investment projects to the design of an efficient and sustainable transportation policy including safety, environmental, accessibility and regional impacts of rail and road investment, maintenance measures and other policy and regulation actions (Widlert, 2001).

The Danish national model PETRA was recently developed as an activity based approach to travel demand analysis (Fosgerau, 2001). The model can be seen as an extension of a trip based frequency model to include an endogenous treatment of the choice of activities in terms of chains of tours. The model has been applied in a variety of analyses, e.g. impact analysis of how different individuals are affected by policy measures. The model development was funded by the Ministry of Transport and by Transport and Energy Research Programmes.

In Italy two parallel efforts to develop national model systems have taken place during the middle of the 90s (Russo, 2001). A decision support system (SASM) for tactical decisions on rail services has been developed by the Italian Railway Company. The system is particularly useful in the analysis of the current situation and in the definition phase of marketing strategies. The second Italian national model system has been developed for strategic transportation policy analysis and has been supported by the Ministry of Transportation and by the National Research Council.

National traffic forecasting in Britain has developed from a partial use of disaggregate models (car ownership) in the 70s, over aggregate traffic growth models in the 80s, towards more elements of behavioural modelling and treatment of supply constraints in the National Road Traffic Forecasts of 1997 (Gunn, 2001; Worsley and Harris, 2001a). Further model developments are envisaged for greater policy sensitivity. These developments may eventually lead to an integrated national transport policy model.

Within the 4th Framework Programme of the European Union modelling efforts have been pursued at various spatial levels. Two such efforts at the European level will be covered in our analysis: the STREAMS model for passenger and freight movements (Williams, 2001) and the passenger demand part of the STEMM project (Gaudry, 2001). Also some of the national models cover transport across national borders, e.g. the SAMPERS model of Sweden and the Italian strategic decision support system.

1.3 A Comparative Analysis

In this section a comparison will be made of the models presented in subsequent chapters of this book: the Netherlands National Model System (NMS), the Tactical and Strategic Decision Support Systems (DSSs) for Italy, the PETRA model for Denmark, the SAMPERS national model system for Sweden, the National Road Traffic Forecasts (NRTF) model for Great Britain, the STREAMS model for Europe and finally the Multicountry Application for Passengers (MAP-1) within the European STEMM model.

1.3.1 Model Structure and Theoretical Approaches

The oldest system represented in this review is the Dutch NMS from the mid 1980s. It extended earlier urban/regional models to the national level. Age-cohort-based licence holding and car ownership models are linked to tour frequency, mode and destination choice models. The resulting O/D rail passenger and car driver trip matrices are assigned to the rail and road networks – in the latter case choosing the time of the day in response to congestion levels on the roads. These models are highly disaggregate, usually based on individual utility maximisation represented in the form of multinomial nested logit models. When applied to forecasting, enumeration of prototypical samples are used together with the 'pivot-point' approach for driver and train passenger flows, i.e. the model system is only used to calculate changes that are applied to 'observed' base-year O/D matrices (Gunn, 2001; Hofman, 2001).

The Dutch NMS is now being updated. According to Daly (2001) the logit structure of utility maximisation leads to a clarity of the behavioural mechanisms that makes a model more easy to update. He also strongly recommends linear formulations of variables whenever this admits behaviour to be reasonably well represented.

Also the Italian DSSs rely heavily on disaggregate nested logit models. As compared to the Dutch NMS much more emphasis has been put on the information systems supporting non-specialist users (Gunn, 2001).

The Tactical DSS is for operational planning of rail services. Of special interest is the 'diachronic network' model, which allows the timetables of rail services to be represented so that desired departure/arrival times can be accounted for. The demand sub-models for train services represent for different market segments the choice of high/low speed train, service class and specific run in relation to desired departure/arrival time by means of nested logit models (Russo, 2001). This approach is similar in scope with – but technically different from – the new timetable based assignment module presented in Constantin et al. (2001).

The Strategic DDS is an integrated system for transport policy analysis covering both passenger and freight. For the freight model system regional input-output models are combined with logit models for mode and route choices. Recent development efforts include the estimation of a domestic mode choice model based on logistical principles (to enable the trade-off between size and frequency to be modelled) and a mode choice model for international freight derived from behavioural principles (Russo, 2001).

The Danish PETRA model is very innovative in its attempt to be a kind of activity-bâsed model. Instead of modelling the travel in terms of separate trips, or even tours, a person's total travel during a day, is represented in form of what is termed a chain. In order to reduce the complexities of observed travel patterns, they are transformed into a limited number of chain types (13 including staying at home). An allowed chain may have at most three purposes (activities), carried out during at most two tours, each visiting at most two destinations (Fosgerau, 2001).

PETRA comprises models to handle cohort effects on license holding and car ownership. Conditional on car availability and each tour type occurring in a chain, mode and destination choice submodels of the nested logit type are estimated. Finally, a model of the choice of chain type is specified, where the accessibility to different destinations and modes are measured by logsums from the mode/destination submodels. The system is applied to a model population using sample enumeration. PETRA does not include any network assignment module and hence congestion is not considered explicitly.

The new Swedish SAMPERS system belongs very much to the mainstream tradition. As for the choice models, which handle the traditional choices, it relies almost totally on the nested logit approach. It covers all trip purposes and all trip lengths including international trips. It is a very large system with about 8,500 zones for local and regional trips, 700 zones for domestic long distance trips and in addition almost 200 zones for international trips (Beser and Algers, 2001; Widlert, 2001).

Like the Italian DSSs, SAMPERS is very ambitious in respect of user-friendliness, prompted partly by the mere size of the system. Special software modules have been developed for the design of scenarios, with an automatic control of input data, and for the analysis, aggregation and visualisation of the results (Johansson Sveder, 2001).

SAMPERS uses EMME/2 for the assignment of car and public transport trips to the different networks. Of special interest is the (probably) first large-scale application of the new module for deterministic timetable public transport assignment (see Section 1.3.2 and Constantin et al., 2001).

The recent British traffic forecast tradition has been quite different from what we have discussed so far. As a consequence of bad experiences from the development of a Regional Highway Traffic/Transport Model in the 1970s, top-down rather than bottom-up disaggregate approaches have been predominant since then. Instead of having a comprehensive model system that produces trip matrices by modes that are assigned to detailed networks, the focus has been on overall vehicle kilometres and on their dependence on a few crucial determinants such as GDP, fuel price and demographic factors (Worsley and Harris, 2001b).

The approach has been criticised for not taking e.g. the limited capacity of the roads into account. In the present NRTF model, following a six-year program of research and development, several improvements have been incorporated, still, however, within the same top-down structure. A significant household segmentation is now allowed for. Forecasts of unconstrained or background growth are derived from car ownership and car use and are then confronted with the actual capacity of the roads in a model process that attempts to reflect how increased congestion will redistribute the traffic and reduce the growth (Worsley and Harris, 2001a).

With the STREAMS model the geographical scale is enlarged to the EU level. On the demand side both passenger and freight movements within the EU are modelled in a highly segmented way using a zonal system of about 200 zones. Both short and long distance trips with all modes are included, which for this scale means a considerable amount of intra-zonal travel requiring a special treatment of local trips including walk trips. Again the logit form is applied and trips and freight transport are assigned to the networks for the different modes (Williams, 2001).

In developing a comprehensive model one can distinguish between an empirical approach, where the emphasis is on a detailed representation of origin-destination modal demand data, and a behavioural approach, where the main interest is on understanding the determinants of the travel demand. The first one may give more accurate traffic levels, especially in the short run and for specific localities. The second one may be preferable when studying long-term effects of factors that may influence behaviour. Special techniques for combining these approaches have been developed (incremental logit, residual disutility) (Bates et al. 1987; Williams and Beardwood, 1993).

Also the MAP-1 model within the STEMM passenger demand project operates at the European level. It pertains to three trip purposes (business, private, vacation) and combines disaggregate mode choice models and aggregate total demand models within a quasi-direct format structure.

The model approach employed in that study challenges the conventional approaches, such as the multinomial logit approach, by relaxing many of the restrictions that are associated with these approaches. This is the case with the homoskedasticity of the unexplained, random, part of the utilities in the logit model. By allowing for direct and inverse Box-Cox and Box-Tukey transformations much more general specifications of the utilities than usually are tried can be estimated. This has important consequences since in a Box-Cox (non-linear) logit model this allows the impact of a network improvement to depend on the level of the characteristic, and also values of time to vary both across modes and with the amount of time saved.

In addition a method for addressing the issue of spatial correlation of the residuals in e.g. generation-distribution models is presented. This can be used for detecting the presence of spatial competition (Gaudry, 2001; Gaudry et al., 1998).

1.3.2 Timetable Based Public Transport Assignment

Network assignment of public transport trips is in applied large-scale transport models typically based on average headway. This may be a reasonably valid assumption for urban public transport systems with high frequency of services. For long distance trips, and in rural areas with low frequency of services, the attracted demand will be very dependent on how well the timetable actually fits the desires of potential passengers. For national transport models timetable based public transport assignment is therefore a very important issue.

Among the national transport models discussed in this paper, only the Italian DSSs and the Swedish SAMPERS system have included such an option. In the Italian case a stochastic network assignment procedure for train trips, based on the 'diachronic network' concept (see Section 1.3.1), has been specifically developed for that system. SAMPERS, on the other hand, has chosen to include the newly released deterministic timetable transit assignment module in the EMME/2 software package.

We will briefly review this new algorithm of EMME/2 (for a detailed description, see Constantin et al., 2001).

The basic problem is to find the best choice of a path for a public transport passenger, given an origin node, a destination node, information on desired departure or arrival time and on the timetable of each relevant line. What is the best choice is determined by calculating the total cost for each considered path by weighing the individual travel time components and boarding penalties that are associated with it.

The timetable information is specified via a number of attributes with the following meaning:

– the *headway* is the interval between successive runs of a line,

- the *offset* is the departure time of the first run after the beginning of the period,
- the *number of runs* is the number of times a line operates,
- the *segment time* is the travel time on a transit segment, and
- the *dwell time* is the time spent at a node serving passengers.

The stops of a particular run are easily determined from this information.

To be able to select the optimal path for a passenger the desired departure or arrival time has to be specified. This can be done in form of a target time and the maximum earliness and lateness with respect to the desired target time. In this way a range of allowed departure (arrival) times is defined. When evaluating a specific departure (arrival) time, penalty weights can be applied for the positive or negative deviation from the target time.

Given the range of allowed departure (arrival) times, which determines the set of feasible space-time paths for the passenger, the path with the least total cost is selected. This is done by applying a least cost path algorithm on a space-time network, given by the timetable information plus the transit and auxiliary transit (walk) network. This network is typically very big and it would be very costly to represent it in its entirety. This is avoided, however, by using an algorithm that generates dynamically only the part of the network that is actually needed for the computations. To reduce the number of alternatives to be explored when building up the feasible paths, an ingenious use of dominance relationships is applied.

After the assignment has been completed the following results can be obtained for each O/D pair: individual travel time components (early or late time, auxiliary transit time, waiting time, in-vehicle time), individual transit cost components (early or late cost, auxiliary transit cost, waiting/boarding cost, in-vehicle cost), number of boardings and distance travelled.

1.3.3 Data and Model Estimation

New data sources and estimation methods have stimulated some of the recent developments of national and European transport models. Lack of appropriate data has also affected the modelling approach in some cases (Britain and EU) but with quite different results. The assumptions on the functional form of travel demand are constrained by the availability of easy-to-use estimation methods. In this section we will briefly comment on data and estimation methods used for the various national and European modelling efforts. We will also touch upon a discussion on the implications of data access for the choice of modelling approach and on estimation methods and functional form.

Most of the modelling efforts rely on a combination of household travel survey data and specific traveller surveys (e.g. onboard and border crossings interviews, mailed interviews to car drivers). The size of household travel surveys ranges from 13,000 used for estimation of the Danish PETRA model to 68,000 used in the recent re-estimation of the Dutch national model (the earlier NMS model was estimated on a small regional household travel survey with only 1,000 observations complemented by about 4,000 rail and road traveller surveys). Also for the European transport models national personal travel surveys constitute the main data sources. However, the information on cross-border trips is limited. Gaudry (2001) uses French and British surveys for multicountry applications of approaches that are first validated on national data.

The number of zones ranges between 200 (NUTS-2, European STREAMS model) to 8,500 (regional models in the Swedish SAMPERS system). The re-estimated Dutch national model is based on a network with about 1,300 zones and the European STEMM demand model is estimated on about 1,000 zones (NUTS-3). All the models except the British NRTF model use data from network representations, even though the Danish PETRA model and the discussion of the European STEMM trip demand model do not contain explicit references to any network model. Specific models for traffic assignment are included in the Dutch, the Swedish and the Italian national model systems and in the European STREAMS model system.

Only two of the model systems do cover freight explicitly: the Italian strategic DSS and the European STREAMS model. Both models derive freight matrices from regional economic approaches and use available freight data for calibration/validation. In the British NRTF model freight traffic developments for different types of vehicles are related to GDP growth.

The information on estimation methodology and software is very scarce in the short reports on national and European transport models that follows in subsequent chapters. Maximum likelihood estimation of linear models is the general approach. Gaudry (2001) uses the estimation software TRIO for endogenous choice of functional form by means of Box-Cox and Box-Tukey transformations.

The Dutch and the Swedish model systems represent similar enhancements as compared to the previous generation of models. However, while the SAMPERS model system is considered as a new development, the Dutch model system is described as re-estimated and related to updated base matrices. In the Dutch case, a series of models is estimated with increasing differences from the old model system: use existing model with new data, use only aggregate data for re-estimation, etc. These tests will illustrate how improvements in the sophistication of the re-estimation and updating procedure improve the quality of the model (Daly, 2001).

The methodology for estimating (or updating) the Dutch base matrices is of general interest, since many of the model systems are used for projecting changes in demand which are applied to the most recently available base matrices.

1.3.4 Validation

There is no doubt that today's state-of-the-art transport model systems are very sophisticated, not the least in respect of the mathematics and computer skills involved, are theoretically well founded, often use advanced estimation techniques to extract maximum amount of information from the data sources available, and are capable of handling many more policy relevant behavioural mechanisms than earlier models. Nevertheless it is still not possible to get a scientific answer to the simple question: How large is the error in a forecast produced by such a system? The reason is partly that this is a difficult question to answer, partly that too little effort has been devoted to the development of a methodology for validating large-scale transport model systems.

One step towards such a methodology could be to develop a kind of scheme, or checklist, for analysing how well a transport model system functions and serves the purposes for which it has been designed:

- *Practical validation (system design and scope of application).* What is the system level? Which parts of the overall transportation system are included? Which mechanisms are endogenous and which are exogenous? What kind of resources (data, money, computers, competencies) have been and will be available? What kind of policy issues are or should be amenable to analysis? Is the system designed in a way that minimises the risk of unintentional misuse (data checks, self-documentation, transparency, user-friendliness)?
- *Theoretical validation.* What is the theoretical foundation? Does the system use an equilibrium or a dynamic approach? How are different submodels coupled to each other? Are the various causal relationships reasonably well modelled?
- *Internal validation.* How good are the models at reproducing the data on which they have been estimated (goodness of fit)? Are parameters of right sign and statistically significant? Is the responsiveness to changes in explanatory variables reasonable (sensitivity test)?
- *External validation.* Can the model system reasonably well reproduce other independent data (such as traffic counts if these data are not used for calibration)? Are the elasticities of the model system in accordance with what can be found in the literature? How well can the model system reproduce a future year (forecasting) or a previous year (backcasting)?

To carry out a validation study along these lines requires a very good documentation of the model system. Historically, a lot of sins have been committed in this respect. The model developer has often been more interested in improving the model itself than documenting the current version.

Among the model systems reviewed in this paper the Italian DSSs and the Swedish SAMPERS system put considerable effort on providing a user-friendly surrounding software support. They are both of recent date and may represent higher demand in this respect from the user's side. Computer software that helps the user to avoid mistakes in specifying model parameters and scenarios can be very important to improve the quality of a particular policy study.

As has been pointed out by e.g. Daly (2001), a successful model development may stimulate applications that go far beyond the initial intentions. Still it is important to clarify the scope of application before a model development is started. There is a trade-off between the time horizon of the planning decision or policy analysis and the level of detail. Lee (1994), in his discussion of this, distinguishes between strategic, tactical and implementation planning. These different planning levels may all need different tools. The ambition to create a comprehensive tool, as is the case of SAMPERS (Widlert, 2001), may turn out to be difficult to accomplish.

Modelling trip chains, as in the PETRA model, seems to be an interesting extension of modelling practice from a theoretical perspective. With respect to internal validation, however, it is not so simple to decide whether this innovation has actually improved the performance of the model system or not. Fosgerau (2001) expresses some doubts whether the low fuel price elasticities he finds really could be trusted.

It is now common practice to use sophisticated statistical techniques to estimate the parameters in the models. The progress has been very rapid. This means that the internal validation is often carefully done, usually indicating well validated models. But there is still room for improvement. As an example, when specifying logit models available software often requires the utilities to be linear in parameters. As is indicated by Gaudry (2001), improved models can be estimated by allowing a more flexible utility structure. Furthermore, the homoskedasticity assumption of standard logit models has seldom empirical support. It is rather applied because of practical computational reasons.

In analysis of travel behaviour certain stabilities are often observed. As one example, the time allocated to travelling purposes appears to be fairly constant over the years and among different categories. An important consequence for travel forecasts is that investments leading to increased speeds tend to result in longer trips rather than reduced travel times. Some researchers even argue that this observation is the outcome of a general behavioural law (for a critical discussion of this position, see Höjer and Mattsson, 2000). An alternative explanation is that travel behaviour is subject to monetary and, in particular, time budget constraints. When estimating disaggregate travel demand models, where different trip purposes are modelled separately, it is difficult to apply such budget constraints in a proper way (one approach is discussed in Eliasson and Mattsson, 2000). Without doing it, there is a risk that the resulting model system might give rise to forecasts of travel demand that are not consistent with such aggregate restrictions. At a more general level, there is a need for an improved estimation methodology that allows disaggregate estimation procedures to account for aggregate restrictions, be it budget constraints, constraints on trip production or attraction or even restrictions on the elasticities of the whole travel demand system.

Another problem related to the external validation is how estimated model parameters possibly should be changed when the model system is used for producing forecasts. For example, the quotient between the travel time and travel cost parameters determines the value of time in a model. It is reasonable to believe that the value of time will tend to increase as income increases through time. This calls for a change of the time and cost parameters when the model system is used for making forecasts. However, there are infinitely many ways of changing the parameters to achieve a certain value of time, and which one we choose will have implications for the policy responsiveness of the model system. A well founded methodology for how this parameter adjustment should be carried out is therefore needed (cf. Williams, 2001). Some interesting research efforts along these lines are reported in Gunn (2001).

The external validation is the really demanding task. Few, if any, national transport model systems have been validated in this sense. However, Gunn (2001) makes many interesting observations in his account of the historical development of such models. Lack of data has often prevented the developer form comparing model forecasts with observed development over a longer time period. Backcasting could be a way of avoiding the lack of data and has also been considered, but dismissed, for the Dutch NMS. In any way, the ability of a model to predict the dynamics of the system modelled is the ultimate test of how good the model is.

Gunn (2001) discusses the methodology of such evaluations of the long-run 'forecasting' accuracy in relation to the Dutch NMS. He also presents some comparisons between forecasts and actual outturn. One observation is that it may be interesting to separate discrepancies between forecasts and actual development due to differences between assumed and actual levels of exogenous key input variables from those that can be related to deficiencies in the model framework itself. Good before-and-after databases related to major transport system changes are required for ultimate model validation. An effort to build such a database related to the Öresund link and its potential use for model validation is discussed in Lundqvist (2001).

In addition to an evaluation of the forecasting (or backcasting) accuracy, a spatial comparison of model systems could be carried out, i.e. a test of their transferability. Following the idea of the ISGLUTI study (Webster et al., 1988) different models could be applied to one spatial setting or one model to different spatial settings. Although this is not a simple task, it is surprising that so little of systematic spatial comparison has been attempted.

1.3.5 Areas of Application

While some of the presented models are of a research character with no specific applications reported or are recently developed and not yet applied, it is clear that the overall impression is that an increased richness of applications and new demands for policy sensitivity have been some of the main driving forces behind the recent developments. The initial objectives for the development of the models differ:

- harmonisation of regional planning efforts;
- preservation of land for strategic transport links;
- evaluation of the efficiency of policies;
- assistance in formulation of planning and management policies;
- spatial impacts of socio-economic and transport policy changes; and
- strategic environmental assessments.

Examples of actual applications are reported in the relevant chapters on each model. The British NRTF and the Dutch NMS models have a long and successful history and hence represent a wide and elaborate range of application areas. Most of our attention in this section will relate to these model systems, which have developed gradually in dialogue with applications over a 10-20 year period.

The NRTF model was initially mainly used for producing traffic forecasts to cost-benefit appraisal and environmental impact assessments of road construction schemes in England. An interesting feature of these forecasts was that they were presented as ranges related to uncertainties in the underlying economic growth and fuel price assumptions. The model was not seen as a tool for designing national transport policies. In the late 1990s the policy context changed and, after certain model developments, the NRTF model was used for providing information on the likely impacts of congestion charging and on the impacts of alternative fuel tax/price assumptions on forecasts of road traffic's contribution to CO_2 emissions. The increasing demand for policy sensitivity has initiated further developments of the model in order to be used for setting national traffic growth targets and air quality targets.

The Dutch NMS model was initially developed for evaluating policy scenarios and as a tool in preparing the Second Netherlands Transport Structure Plan. The model system has subsequently been used e.g. for assessments of rail options, for defining scenarios for use in policy evaluation and for evaluation of related policy scenarios. While certain applications have demanded development of related additional models (regional models; detailed choices in using the rail mode), many new applications have been possible to accommodate by extending the NMS model (e.g. for analysing road user charges (time switching; highway toll passes), parking restraints, impacts of 'telematics', traffic management measures to maximise the use of the present highway capacity and the impact of increasing fuel prices with due regard to fuel efficiency). Recent applications show a shift in emphasis from environmental concerns to accessibility consequences of policy options.

The Italian decision support systems (SASM (rail DSS) and SISD (strategic DSS)) have been developed for operational planning of rail services and for analysing strategic transport policies, respectively. SASM is used by the Italian State Railways for analysing corridor marketing plans and strategies. The decision variables include operational characteristics (paths, stops, timetables and train compositions). The SISD is used by the Ministry of Transport for monitoring of the Italian transport system, for assessment of policies and scenarios related to the Italian Transportation Master Plan and for supplying information to operators.

The Danish PETRA model has typically been applied to incidence analyses of how different individuals are affected by policy measures, e.g. price and service elasticities for public transport in various urban areas, demand for long distance busses, effects of income tax deductions on commuting, and mobility and taxation differences between urban and rural areas. In Fosgerau (2001) the effects of fuel price increases in the base scenario and of changing household structure and income increases in a future scenario are analysed.

The new SAMPERS model system in Sweden is intended to be used over a long time period and with further developments of the modular structure over time. The main application of the model system will be in strategic investment planning, but evaluation of transport policy measures have been more closely integrated with the evaluation of infrastructure investments during recent years. This leads to widening application areas: in addition to impacts of new infrastructure projects or programmes in different socio-economic settings there are also needs for evaluation of transport policy goals and evaluation of effects of transportation policy measures (e.g. taxation, pricing, administrative rules etc.).

Finally, the growing importance of environmental impact assessments (EIA) in applications of national model systems needs to be emphasised. EIA is now compulsory in project appraisal in most countries. Strategic environmental assessments (SEA) are being introduced for early evaluation of environmental consequences of policies, plans and programmes. A decision on a SEA Directive within the European Union is scheduled for early summer 2001.

1.4 National Models in Perspective

In this concluding section we will comment on some general characteristics of national transport models that may be of particular interest for the future development of such models.

In discussing strengths and weaknesses of the Dutch NMS, Hofman (2001) brought up some issues of general interest:

- the majority of models have *comprehensive* ambitions at the national level, they tend to be *too aggregate for regional applications*. In the recent Swedish model system SAMPERS, both national models and more detailed regional models are integrated within a consistent (but very resource consuming) framework;
- some of the models have been used extensively over a long time period (e.g. the Dutch NMS, the British NRTF and the earlier Swedish national models) and have turned out to be *reliable* tools. Nevertheless, a number of aspects can be listed where further improvements are desirable: development of *activity based approaches* for treatment of space-time relationships between various activities and trip types, and introducing *dynamics of transport behaviour* (e.g. habit formation and anticipations and timing of behavioural changes); and
- in spite of the reliability and good experiences from some national models in their respective policy contexts, increasing demands for *policy sensitivity* have been the main driving forces behind new model developments and flexibility in model design has been emphasised in order to meet new policy demands. Similar demand for increased policy sensitivity has also been reported from the U.S. (Lakshmanan, 1998).

All models in this study assume that regional development or land-use data are exogenously given as part of input scenarios. Hence, regional development and land-use are not assumed to respond to changing conditions in the national transport systems. This may be an appropriate assumption in cases of predominantly non-congested networks. However, in most European countries investments have not matched the traffic growth during recent decades which has lead to increasing levels of congestion. Hence, *interactions between the transport systems on the one side and regional development and land-use patterns on the other side* tend to be more important and requires integrated approaches for modelling of two-way interactions. This is further underlined by increased demand for decision support systems capable of providing information on the long-term effects of infrastructure investments for strategic environmental assessments (Wegener, 1998).

Prevailing national transport models incorporate iterations between transport demand and transport supply (traffic assignment) at best. Some of the models are so complex that even iterations might be cumbersome. Again, the models might perform satisfactory in cases where the level of congestion is low. Changes in demand would in such cases not affect the conditions in the transport systems very much and the need for feedback iterations would be small. For the same reasons as above, increasing congestion levels may lead to a new emphasis on *integration of demand and supply*. Hofman (2001) has indicated some problems in finding such an equilibrium by feedback iterations. Experiences from combined network equilibrium models (see e.g. Abrahamsson and Lundqvist, 1999) might be used in order to ensure convergence and increase the speed of convergence.

Most models represented in this volume (e.g. the Dutch NMS) are of the '*bottom-up*' type: models are estimated for individual travel demand based on disaggregate data on travel patterns and transport networks. One example of the '*top-down*' approach is the British NRTF model with aggregate travel demand forecasts without explicit trip matrices or spatial transport network representation. A discussion of conditions where any of these approaches might be preferable is provided in Worsley and Harris (2001b). It should be noted that both the Dutch NMS bottom-up model and the British NRTF top-down model have been used successfully during a long time period.

There are arguments for both linear and non-linear models in this volume. In Daly (2001) it is argued that flexibility in applications requires simple models with standard *linear utility forms*. Even if an accurate representation of behaviour might suggest more complicated model forms, this might reduce the possibilities for applying the model widely and reliably. On the other hand Gaudry (2001) argues for generalised *non-linear model forms*. The Box-Cox form has long demonstrated its ability to provide better fits to observed data and more reasonable results than the linear model form. Also the applications on European data would reject the multiplicative, gravity-like model specification including the logsum aggregator. These arguments would indicate a trade-off between ease of applications and accurate representation of behaviour.

As noted above a few national models (both bottom-up and top-down approaches) have exhibited a long and healthy life time. Daly (2001) discusses various ways to extend *the life time of a model* in response to new conditions (e.g. new policies or market changes) by collection of new data, by careful interpretation of policies or by an adaptive model design (based on simple and transparent model forms). The discussion is related to the bottom-up NMS model. Similarly, Worsley and Harris (2001b) argue that the top-down model approach, as represented by the British NRTF model, lends itself easily to an evolutionary development. Both the NMS and NRTF models are subject to substantial current improvements. The border-line between what might be called model improvement and what might be called development of a new model is not very clearly defined.

Our final perspective relates to the *user friendliness* of national transport models. Very few contributions discuss this point, which is surprising with regard to the rapid development of computer software in general, and *geographical information systems (GIS)* in particular, during the 90s. Only in the descriptions of the new Swedish model system SAMPERS (Johansson Sveder, 2001; Widlert, 2001; Beser and Algers, 2001) the importance of a user-friendly integrated system is explicitly taken up. The emphasis is on the user interface for scenario formulation and presentation of results. Later some mapping functionality has been added for presentation of accessibility results. However, there seems to be no ambition, neither in the Swedish case nor in any other national model, to achieve a close integration between any particular GIS and the national transport model system.

References

Abrahamsson, T. and Lundqvist, L. (1999), 'Formulation and estimation of combined network equilibrium models with applications to Stockholm', *Transportation Science*, vol. 33, pp. 80-100.

Bates, J., Ashley, D. and Hyman, G. (1987), 'The nested incremental logit model: Theory and application to modal choice', in *Proceedings of the PTRC Summer Annual Conference, Series C*, PTRC Education and Research Services, London, pp. 75-82.

Beser, M. and Algers, S. (2001), 'SAMPERS – The New Swedish National Travel Demand Forecasting Tool', Chapter 9, this volume.

Constantin, I., Florian, M. and Spiess, H. (2001), 'Deterministic time table transit assignment', Chapter 6, this volume.

Daly, A. (2001), 'Updating and extending national models', Chapter 3, this volume.

Eliasson, J. and Mattsson, L.-G. (2000), 'A model for integrated analysis of household location and travel choices', *Transportation Research A*, vol. 34, pp. 375-394.

Fosgerau, M. (2001), 'PETRA – An activity-based approach to travel demand analysis', Chapter 11, this volume.

Gaudry, M. (2001), 'Tests of nonlinearity, modal captivity and spatial competition within the STEMM multicountry application for passengers', Chapter 14, this volume.

Gaudry, M., Heinitz, F., Last, J. and Mandel, B. (1998), 'Methodological development within the quasi-direct format demand structure: The Multicountry Application for Passengers MAP-1', Bureau d'économie théorique et appliquée, Working Paper No 9815 and Centre de recherche sur les transports, Publication CRT-98-46.

Gunn, H. (2001), 'An overview of European national models', Chapter 2, this volume.

HCG and TOI (1990), 'A model system to predict fuel use and emissions from private travel in Norway from 1985 to 2025', Report to the Norwegian Ministry of Transport, HCG and TOI.

Hofman, F. (2001), 'Application areas for the Dutch national model', Chapter 7, this volume.

Höjer, M. and Mattsson, L.-G. (2000), 'Determinism and backcasting in future studies', *Futures*, vol. 32, pp. 613-634.

Johansson Sveder, G. (2001), 'The new Swedish national model SAMPERS: System and validation', Chapter 8, this volume.

Lakshmanan, T.R. (1998), 'The changing context of transportation modeling: Implications of the new economy, intermodalism and the drive for environmental quality', in Lundqvist, L., Mattsson, L.-G. and Kim, T.J. (eds), *Network Infrastructure and the Urban Environment*, Springer-Verlag, Berlin, pp. 53-71.

Lee, D.B., (1994), 'Retrospective on large-scale urban models', *Journal of the American Planning Association*, vol. 60, pp. 35-40.

Lundqvist, L. (2001), 'A before-and-after database for land-use/transport/environment impact analysis and model validation', forthcoming in Karlström, A. and Kaag Andersen, A. (eds), *The Regional Development Consequences of Transport Infrastructure Investments - With Special Emphasis on the Öresund Bridge*. Research Report, Department of Infrastructure and Planning, Royal Institute of Technology.

Russo, F. (2001), 'Italian models: Application and planned development', Chapter 10, this volume.

Webster, F.V., Bly, P.H. and Paulley, N.J. (eds) (1988), *Urban Land-Use and Transport Interaction: Policies and Models*, Report of the International Study Group on Land-Use/Transport Interaction (ISGLUTI), Avebury, Aldershot.

Wegener, M. (1998), 'Applied models of urban land use, transport and the environment: State of the art and future developments', in Lundqvist, L., Mattsson, L.-G. and Kim, T.J. (eds), *Network Infrastructure and the Urban Environment*, Springer-Verlag, Berlin, pp. 245-267.

Widlert, S. (2001), 'National models: How to make it happen. The case of the Swedish national model system: SAMPERS', Chapter 4, this volume.

Williams, I. (2001), 'Designing the STREAMS model of Europe', Chapter 13, this volume.

Williams, I.N. and Beardwood, J.E. (1993), 'A residual disutility based approach to incremental transport models', in *Proceedings of the PTRC Summer Annual Conference, Series D*, PTRC Education and Research Services, London, pp. 11-22.

Worsley, T.E. and Harris, R.C.E. (2001a), 'GB traffic forecasts – Status and development', Chapter 12, this volume.

Worsley, T.E. and Harris, R.C.E. (2001b), 'General modelling approaches: Top-down or bottom-up?', Chapter 5, this volume.

Part I

The Model Development Context

2 An Overview of European National Models

Hugh Gunn
RAND Europe
Newtonweg 1
2333 CP Leiden, The Netherlands

2.1 Introduction

The aim of this chapter is to provide something of a historical perspective from which to view recent advances in National Travel Demand Models, as an introduction to the discussion in other chapters of this book.

In 1997 a seminar on National Travel Demand Models was held in the Netherlands that was attended by some forty delegates from eleven European countries. At that event, information on the state-of-the-art in country-wide modelling was provided for Belgium, Denmark, Germany, Hungary, Italy, the Netherlands, Norway, Sweden and the U.K.

Of these countries, four were at the stage of considering possibilities for a new, large-scale national model (Belgium, Denmark, Sweden and the U.K.), two reported on existing system based almost exclusively on traditional aggregate four-stage models (Germany, Hungary), and three which reported on recent disaggregate national models (Italy, the Netherlands and Norway).

The major purpose of the Seminar was to establish a forum at which those actively involved in European National Modelling, clients, researchers or consultants, could meet to exchange views and findings.

In the aftermath of that discussion, the focus of this chapter is on the emerging stream of *disaggregate* model structures over the last twenty years; in turn it looks at

- the first *partial* use of disaggregate models at a national level in the Regional Highway Traffic Models (abbreviated RHTM) for England (commenced 1976),
- the first *comprehensive* country-wide application in the Dutch National Model System (commenced 1983),
- the extension of that system to create the Norwegian National Model (commenced 1989), and finally
- the later developments in the Italian National Model (commenced 1992).

For each of these national models, we look first at

- the objectives of the modelling study,

- the broad structure of the model system,
- the data base,
- the scale of analytic effort required,
- validation, and finally
- the major lessons drawn from the work.

2.2 The RHTM Model of England (1976-1978)

2.2.1 Objectives

The principal purpose of the RHTM project was to provide a unifying modelling framework (including values of coefficients of component models and levels of strategic origin-destination matrices for base and forecast years) to harmonise the planning and construction of road schemes as between different regional authorities.

The best single reference document to the study is the RHTM Project Review (Alastair Dick and Associates, 1978).

As a consequence of the standardisation that could be imposed by the success of the project, the economic evaluation of the relative merits of Scheme A in Region 1 and Scheme B in Region 2 would not be invalidated by differences in forecasting assumptions, or technical aspects of the models used. In addition, after assignment to a strategic national network, matrices of car-driver trips between external zones and internal zones, and between external zones and other external zones, could be supplied to each region on a consistent basis. This was intended to avoid the possibility of inconsistent assumptions about the routing of long-distance traffic through the different regions.

Another very important consideration in the planning of the work was the damage that was being caused by so-called 'planning blight', a process in which buildings in locations thought possible for road extension could lose virtually all re-sale value for very many years.

The time period for the model was to be an annual average day during the working week. Travel was represented as a series of unlinked trips. No specific time-of-day variation was included in the basic design.

Finally, well worth noting are the first two of the issues stated in the 'Requirements of a Revised Procedure' noted in the RHTM Project Review

- that the procedure should be 'rational, so that it can be readily understood by the public'; and
- that the model should be 'nationally consistent in its approach'.

2.2.2 Structure

The RHTM models retained the approach then described as the traditional 'tried and tested' structure developed for standard urban and regional *mono-modal* models, with models of car-ownership leading to models of car-trip making at a zonal level, and with the resulting trips being distributed between available destinations and then assigned to networks.

The car-ownership model stage, commenced in 1976, reflected some of the thinking that had underpinned disaggregate analyses in the Netherlands during the early 1970's, in that it looked to emphasising *behavioural* aspects of the car-purchasing behaviour as the explanation of variations in levels of car-ownership as between different locations and different time-periods (Bates et al., 1978). This emphasis on behaviour prompted the adoption of disaggregate techniques aimed at individual decision-making units, here taken to be the household, and explanatory factors affecting the decision, here taken to be real income and some measure of cost.

The actual structure of the car-ownership model identified three discrete states of non-car-owning, single car-owning and multiple car-owning, and estimated two models to distinguish between these. These models respectively give the probability of owning at least one car and the probability of owning two or more given that at least one is owned, both as a function of real income and some cost variable. The production of aggregate forecasts, whether zonal or national, involved integrating over an appropriate income distribution (cost being fixed in any year), with zonal models using adjusted coefficients specific to the zone type (Bates et al., 1979).

In fact, the allocation of households to income classes was approximated by a Gamma distribution, resulting in a simple, one-dimensional sample of households distributed over the single explanatory variable, income, which could be used, together with the calibrated models, to predict numbers of cars owned. In later systems, the device for producing forecasts from multivariate models by integration over multidimensional 'prototypical samples' could be seen as an extension of this approach.

Trip frequencies were taken as simple rates within a small number of categories of traveller (conditioned amongst other things by household car-ownership), and trip distribution performed by a small number of gravity models. The assignment stage was never completed.

2.2.3 The Data Base

The RHTM data base involved some 40,000 household interviews and several hundred thousand roadside interviews (involving stopping motorists and filling questionnaires on the spot). The background planning data and networks had also to be assembled from scratch for some 3,600 zones.

The car-ownership model used an additional source of information to try to establish a forecasting capability; this was the Family Expenditure Survey, conducted annually on samples of around 3,500 households. Data from several years was analysed in an attempt to demonstrate a time-stable relationship.

2.2.4 The Analytic Effort

The RHTM project was undertaken as a collaborative venture by three major English consultancies collaborating with local and central government officers. Teams of around ten persons were occupied more-or-less full-time for around two years, at each of six regional sites (with student and casual labour for coding tasks on an occasional basis). Different models were typically each developed at a separate site, and then transferred to the other sites for local validation.

2.2.5 Validation

As mentioned above, the car-ownership model was estimated and validated against time-series data to demonstrate forecasting accuracy. In the time-scale of the project, the model appeared to hold quite well; however, in subsequent years it became apparent that the model was not properly specified leading (amongst other developments) to the use of driving-licence information in the analysis and prediction phases.

Zonal base-year income estimates (expanded with the gamma-function assumption to provide the 'prototypical sample') allowed base-year zonal car-ownership totals to be generated, and these were then validated against an independent source of local car-ownership levels.

With the problem of the time-stability of the predictions yet to emerge, the system appeared reasonably on-course at this stage. However, the next stages of developing the numbers and patterns of movement of trips proved more difficult. Origin-destination matrices were developed by the 'tried and tested' methods, but comparisons of large-area to large-area flows with independent cordon-crossing counts revealed large biases which were never fully explained, except as a demonstration of inherent over-simplifications in the model.

A subsequent in-depth investigation of the problem (Gunn et al., 1980) raised several problems apparent in the data and perhaps inherent in the approach:

1. the dominant role played by the intra-zonal movements, for which no satisfactory measure of 'separation' had been found; and

2. the obvious difference between the household and roadside interviews in terms of their distribution of trip by purpose.

In addition, conjectures were made over the appropriateness of the use of a mono-modal model within a country in which there were large variations in the density and use of public transport.

The lack of an available solution to solve the problem eventually led to the project closing without fulfilling the most important of its original aspirations, although (with later adjustments) the car-ownership models continued in use and the planning data-base was made generally available.

2.2.6 Major Lessons

At a cost of £5,745,000 by 1978 (perhaps mEURO 50 in 1998 currency) the major conclusion that has been drawn by the English planning community seems to have been very negative for national modelling.

However, other constructive lessons can and have been drawn; these include the following:

1. For car-ownership trends consideration of patterns of licence acquisition and disposal can improve forecasts.
2. Even if only one mode is really of interest, all competing modes have to be in the model if current behaviour and future changes are to be realistic.
3. If intra-zonal attraction cannot be modelled adequately, errors can at least be minimised by reducing the size of the zones (i.e. increasing the number).
4. Synthetic models can be very poor predictors of actual O/D patterns. Rather than play this down by incorporating K-factors in a late stage of the model process, it may be more acceptable to present the models as predictors of change to be applied to best-available base matrices (pivot-point), even the end result is similar or even identical.

Less useful during the preparation for inevitable widespread road construction in England in the 1970's, but actually rather interesting in the late 1990's, could have been the conjecture that technological advance might reduce the computing costs by factors of thousands, and models should not be overly restricted by current computational restrictions.

2.3 The Dutch National Model System (1985 – Present)

2.3.1 Objectives

The principal objective of the Dutch National Model System (NMS) was to make provision for the preservation of access for strategic road and rail links yet to be developed. It had to be *national* because it was to be expected that some or all of these links would cross the entire country. At a later stage, it became an important tool in the development of a national environment plan, and also found use as a forecasting tool for the Dutch national railways. A twenty-five year time horizon was originally set, having the reporting year 2010 (DVK, 1990; Gunn, 1994; TNO, 1996).

Travel is represented as a group of home-based tours for the purposes of forecasting growth and change, but as unconnected trips for assignment to networks. The time period for the models was an annual average day during the working week, but an allowance was made to distinguish peak hours from off-peak.

2.3.2 Structure

Like the English model system, NMS was based on an extrapolation of an existing urban/regional system up to the national level. Unlike RHTM, however, this was an entirely disaggregate system, with stages of licence holding, modal choice and time-of-day all linked together with models of car-ownership, trip frequency and distribution, and all based on analyses of individual choice. With a view to increasing behavioural realism, the models dealt with so-called 'primary-destination tours' rather than trips. This followed some experimentation on time-stability, but in brief was intended to focus on the main activity carried out during a tour to the exclusion of detours (many of which involved short stop-overs to and from the primary destination).

Another basic difference was the use of the 'pivot-point' approach to forecasting O/D matrices, using the synthetic models to predict changes to best-available base year matrices for car-driver flows and train flows. For the other modes (which include bus, tram, metro, car-passenger, walk and cycle), the synthetic models can be used to output synthetic matrices if necessary, as an approximation to real flows by these modes.

In the early design, provision was made for a simple model of long-distance business travel to be incorporated; this was an elasticity-based adjustment process applying to a small number of different market segments. However, after comparison with the performance of the standard business models at the long-distance end of the market, this module was withdrawn as providing less-credible forecasts.

A particular feature of the NMS is the use of an age-cohort-based approach to driving-licence acquisition and car-ownership. Derived from a repeated-cross-section National Travel Survey, this approach is used to provide a basis for the forecasting of the response to trends in (e.g.) age, income and occupational status. In this way, it performs something of the same function as the RHTM car-ownership use of the British Family Expenditure Survey, setting aggregate totals for licences and/or cars which are then apportioned between households and zones by more detailed disaggregate models.

2.3.3 Data Base

The NMS mode/destination models were taken from a fairly small regional model system for the Utrecht area. This system was developed on a small (1,000) household survey augmented by around 4,000 surveyed rail and road travellers (either handed self-completion questionnaires on trains, or sent the same through the post on the basis of licence-plate matching with computer records of registered owners).

For national application, the repeated-cross-section household interview data (around 10,000 households per year in the early years) was used to extend the disaggregate frequency, licence-ownership and car-ownership models. Stated-preference surveys were carried out amongst a small number of hundreds of drivers to establish time-of-day switching propensities.

The Netherlands contains around 15 million inhabitants, and was divided into some 345 internal zones (around 1,100 subzones) for the early work.

The system is currently being re-estimated on more recent data; we return to this later.

2.3.4 Analytical Effort

Like the RHTM study, the NMS was originally devised as a collaborative effort between several consultancies. The major stages of car-driver base matrix estimation, freight matrix estimation and overall forecasting model construction all allocated to different consultants, as were the other components of the cohort-based aggregate licence and car-ownership models and the long-distance business travel module.

Production of the first working system took around two years, although subsequent development work has been undertaken in most years, and currently a major re-estimation is underway.

2.3.5 Validation

Back-projection as a means of overall system validation was discussed and dismissed as too expensive and possibly inconclusive at an early stage. Statistical significance of model coefficients, ability to simulate choices for sub-markets of travellers and most importantly system elasticities were taken as the main indicators of a valid system. Some comfort was also drawn from earlier tests of model transferability at the regional level, although it has to be said that these tests were weak, and indicated only 'general reasonableness'.

More recently, sufficient time has elapsed to re-analyse old ten-year forecasts at the national level, attempting to isolate model performance from the accuracy of the input assumptions (see Appendix for a more extensive discussion on this issue). This exercise has drawn generally favourable conclusions for the stability of the system, whilst indicating some specific problems with the exclusion of income from frequency forecasts for recreational trips and also with the treatment of the deterrent effect of travel cost on mode and destination choice.

Since this may be one of the most important areas for taking forward in forthcoming models, it is worth also mentioning the parallel research that has taken place in the Netherlands into time/cost trading. Over the same ten year period, a large-scale Stated Preference experiment has been replicated in almost exact detail, with the intention of isolating unexplained effects over and above the distribution of the travelling population between income, occupation, activity-type and journey-type classes.

The conclusions of this work were that any residual effects were very small (less than 5 % overall) and that the major changes experienced were due to shifts in the income distribution. Re-inforced by compatible studies in England and Wales, the broad conclusion was that the disutility of travel cost diminished with increasing income, but not proportional to increasing income. An elasticity of around -0.5 has been found in both countries. Further work is underway in the Netherlands to check the implications of this finding in backcasting and forecasting with two subsequent versions of the NMS.

In addition, the general reasonableness of the NMS in predicting short-run (under one year or up to five year) changes in urban flows subsequent to the completion of a major urban network link has been checked in Amsterdam, and continues to be monitored.

Validation investigations of the base matrices have not returned such favourable results, however, with very large cordon-crossing and specific link variances in assigned flows. Updated versions of the matrices are currently in production, using techniques first suggested in the aftermath of the RHTM English national model (extended to exploit the detailed disaggregate synthetic matrices available in the Netherlands).

2.3.6 Conclusions

A first, and welcome conclusion of the early NMS work was that a useful national model could be built at reasonable cost. The project had started out with the advantage of (largely) existing networks and land-use data, and in a country with the great advantage of licence-plate identification as a means to distribute a sample of questionnaires to motorists, but at a total cost around mEURO 1.5 in current terms, a forecasting system had been built.

It is fair to restate that this model was never judged on its ability to reproduce base-year trip matrices, or cordon-crossings or link assignments, despite the best efforts of an international Audit to apply this test. The NMS models were designed to extract the impacts of changes in factors which are explicit in the models on a detailed O/D/Mode basis, despite the fact that if no better estimate of a particular O/D/M cell value existed, it could be synthesised by the models.

Later developments are aimed at developing a model system which would pass this test, but not by refining the behavioural models; base-year fit is still seen as a matter for judging base-matrices, and requires additional data (e.g. from counts) to pick up additional local factors not in the general model.

Development work (and expenditure) continues, and it is hoped that more evidence on transferability and specification will be published shortly.

2.4 The Norwegian National Model System (1988–1990)

2.4.1 Objectives

The single major objective of the Norwegian National Model was the predictions of emissions of CO_2 and NO_x on a national scale (HCG and TOI, 1990). Also interesting was the provision of a detailed car-ownership and use model as a substitute for a simpler representation of demand for cars and fuel within an existing computer model of the Norwegian economy (MODAG). Thus, at the outset anyway, detailed link-loadings on an explicit national network were not needed. Further work has taken place subsequently.

The models were designed to apply to an average day in the working week. No peak-hour information was supplied. Travel was represented as linked tours.

2.4.2 Structure

The structure of the system for the prediction of passenger demand followed that of the Dutch NMS virtually exactly, including the focus on tours, with four major changes:

- the addition of a detailed treatment of long-distance travel,
- the omission of individual-zone-specific attributes from short-distance travel,
- the omission of the base-matrix pivoting, and
- the addition of a model of joint car-ownership and use.

The long-distance models were separated out (at 100 km) for convenience, since the level of detail of the characteristics of available modes and destinations were accurate enough for conventional separate treatment, and at any rate the modes of air and sea came into play as available options.

The short-distance models used times and costs imputed from reported distances and relationships between distance and average times and costs taken from extra data available for work trips. The models were used to split tours originating in each zone between available modes and distance-based annuli around the zone, yielding kilometrage forecasts without individual O/D matrices.

The model of joint car-ownership and use was itself a transfer from similar analyses undertaken in the Netherlands and Israel, and was proposed as an option for any forecasting scenarios in which extreme reductions in car-use (through variable costs or fixed costs) were sought in pursuit of environmental policies.

2.4.3 Data Base

The basic data base was the Norwegian National Travel Survey from 1984/5, a one-day diary (plus information on long-distance travel in the previous three months) filled in by over 5,800 individuals residing in Norway; this reduced to around 4,000 after screening. Additional information was taken from an available Budget Survey to give estimates of fixed and variable costs per car and per car-kilometre.

Norway contains approximately 1.5 million households, with zonal 'aggregated' destinations split into 7 distance bands on the basis of the 454 kommunes.

2.4.4 Analytical Effort

The initial analytic effort, split between the two consultancies involved, was only some mEURO 0.25 in current terms, reflecting

- the economies available from using structures based largely on previous research, and
- the savings involved in a non-network analysis.

2.4.5 Validation

The validation reported consisted, as in the case of the early Dutch NMS, of statistical properties of coefficients within ranges (or with signs) suggested by economic theory and previous work, augmented by comparison of key elasticities with those reported elsewhere.

Also as with the early NMS, no validation of O/D matrices, but only aggregate kilometrage within groups of travellers.

2.4.6 Conclusions

Perhaps the principal conclusion from the early Norwegian work is that it is not necessary to work with enormous budgets to develop even quite elaborate simulation models, building on the NMS basic structures. The savings through avoiding network complexities and development costs are marked.

The work included the construction and transfer of application software, allowing the exploration of a full range of price, regulation and speed controls to meet the national environmental objectives. Full forecasts were produced for three forecast years to 2025, and was judged favourably in comparison with the output from the macro-economic MODAG output, particularly in bringing to the process an understanding of the effects of demographic change and the implicit virtual saturation of growth in travel demand per person in wealthy futures.

It is understood that more spatially detailed models have been developed since this work.

2.5 The Italian Decision Support System (1993-1996)

2.5.1 Objectives

The goals of the Italian model system were stated as

- to simulate the behaviour of transportation systems,
- to formulate planning and management policies, and
- to check the effectiveness of proposed policies.

The system, named SISD, is quite spectacular in its ambitions for the information technology around the core models, allowing a very high degree of use by non-specialist (transport, software or systems) users. The core models themselves simulate all travel by all modes between the Italian provinces, and are capable of producing seven-day-week, winter and summer forecasts for both Italian residents and visitors (Cascetta et al., 1996).

Travel was represented as linked tours. Simple methods were proposed to extract time-of-day distribution of travel based on reported outward and inward tour legs.

2.5.2 Structure

The structure of SISD can be described relative to the Dutch NMS as including the following substantial variants and extensions:

- treatment of winter and summer differentiation, in which tourism and recreation are accorded particular attention,
- having a range of assignment procedures bringing together route-choice between identified options with incremental loading to estimate realistic journey conditions, and
- recognising the different hierarchies of intra-regional travel and national travel by pre-loading regional flows prior to considering national routes, this
- in addition to the extensive support-system structures which provide information for policy formulation and assessment.

2.5.3 Analytical Effort

The SISD models were built by Italy's leading IT Consultancy, FINSIEL, co-ordinating other research groups (Government, Service Operator, University and Consultancy) in Italy.

Because of the differences in emphasis placed on user-environments, planning capabilities and synthetic travel demand model construction, it is hard to produce cost estimates which are closely comparable with the English, Dutch and Norwegian experiences. However, a 'guestimate' of around mEURO 2.0 - 3.0 based on data collection plus processing has been proposed.

2.5.4 Data Base

The data base was divided into

- *household-based person records* (telephone elicited) of travel by those over 14 years; 6,000 summer interviews and 10,000 winter interviews were collected,
- *border-crossing records* (interviews at ports, airports and international rail stations), 12,500 in summer and 14,000 in winter, and

- manual traffic counts on 125 main road sections (in both directions, including 13 passes) during a workday and repeated on Sundays at the passes.

267 national zones (circa 1,000 regional) were distinguished, and 62 'foreign' zones. Modes distinguished were: car, bus, air, interregional train, intercity train, sleeper train.

The country contains 56 million people, approximately.

2.5.5 Validation

Normal statistical tests on model parameters and elasticities are available. As yet, no checks on forecasts or backcasts are reported.

2.5.6 Conclusions

The Italian SISD is undoubtedly the most ambitious of the four national modelling projects in terms of the surrounding support software. This is not a focus for this note. However, the complexity of the system structure (in particular the attention to time-of-year variation and the methods chosen to allow assignments on such extensive national networks) seem to be pointers towards valuable extensions of practice. This is so particularly for long, thin countries with climatic extremes and an important, seasonal, tourist industry.

Substantive conclusions have to wait until the system is subject to more testing.

2.6 Conclusions, and Some Points for Discussion

This chapter has attempted to set down a brief history of the development of planning contexts, scientific ideas and computing capabilities which have brought forward the production of four European National Models in the last twenty years.

Firstly, it is hoped that the chapter will provide a kind of reference frame, or background material for discussion of other models (and later versions of the models described here) presented in this book.

In terms of the *purpose* for which National Models have been built, the contexts divide broadly between the various needs to provide *scientific material* to

- co-ordinate regional-level construction efforts (England, the Netherlands and Italy),
- set national-interest pricing and control policies (the Netherlands, Norway and Italy),
- inform international-level environmental debate (the Netherlands, Norway and Italy).

For many Northern European countries, the pre-occupation with strategic construction (TGV excluded) is reducing, but regional and local investments still have to be made. Obviously, the private sector must be informed about the demand that will arise for major links it may be asked to finance, and these tend to involve national flows. Furthermore, for many of the Eastern European countries, construction of strategic links is still a major issue.

The basic problem here is the need for realistic O/D matrices and realistic link flows. Many general policy options can be ranked with only kilometrage, time or expenditure on travel. Is the need for spatial detail diminishing or not? What accuracy is needed? Are we able to produce sufficiently accurate estimates? If not, what is to be done?

In the background, almost independent of these, has been the need to provide *basic information* to

- persuade and inform the planning profession (England, the Netherlands and Italy),
- persuade and inform politicians and public (England and the Netherlands).

Relevant to the first 'need' is that the planning profession seems to be growing; in particular it now embraces 'IT' specialists, through whose abilities the traditional transport modeller must now communicate with the world at large. A good or a bad thing?

Relevant to the second 'need', what emphasis should be placed on the development of good advocacy *as-well-as/instead-of* models which are sufficiently complex to account for the range of factors affecting the forecasts. For example, what proportion of project budget should be spent on reporting, and to whom?

References

Alastair Dick and Associates (1978), 'Regional Highway Traffic Model', U.K. Department of Transport.

Bates, J., Gunn, H.F. and Roberts M. (1978), 'A disaggregate model of household car ownership', Research Report 20, U.K. Department of the Environment.

Bates, J., Gunn, H.F. and Roberts M. (1979), 'A model of household car ownership', *Traffic Engineering and Control.*

Cascetta, E., Nuzzolo, A., Biggiero, L. and Russo, F. (1996), 'Passenger and freight demand models for the Italian transportation system', in Henscher, D., King, J. and Oum, T.H. (eds), *Proceedings of 7th World Conference on Transport Research*, Pergamon, p. 473.

DVK, Rijkswaterstaat (1990), 'The national model for traffic and transport', Report B: Main Points.

HCG and TOI (1990), 'A model system to predict fuel use and emissions from private travel in Norway from 1985 to 2025', Report to the Norwegian Ministry of Transport, HCG and TOI.

Gunn, H.F. (1994), 'The Netherlands National Model: A review of seven years of application', *International Transactions in Operational Research*, vol. 1, no 2, pp. 125-133.

Gunn, H.F. and van der Hoorn, A.I.J.M. (1998), 'The predictive power of operational demand models', presented to *the 26th European Transport Conference,* Loughborough University.

Gunn, H.F., Kirby, H.R., Murchland, J.D. and Whittaker, J.C. (1980), 'The R.H.T.M. trip distribution investigation', U.K. Department of Transport.

TNO Inro (1996), 'Audit of the Dutch National Model: Final report', INRO-VVG 1996-29.

van der Waard, T. (1990), 'Handbook of elasticities', AVV, Rijkswaterstaat.

Appendix: A Validation of the Dutch National Model System Using Ten-Year Forecasts

Introduction

A National Model System (NMS) has been in use in the Netherlands since 1985, focussing on land-based travel demand and particularly on personal travel. The model system has been extensively extended and refined in the intervening years. However, a record of the assumptions and results from the 1985 projections was retained, and there are now published aggregate statistics against which to assess these (Gunn and van der Hoorn, 1998).

This appendix provides an introduction to that work. The NMS is a large and continuously developing model system; however, the core of the model remains unchanged, which is to say structure, logic, and indeed the actual utility-maximising models in the joint mode-destination sub-model.

In terms of an understanding of the *basis concepts* of the of the model system (prototypical samples, cascading choice hierarchies etc.) and the *underlying principles* of its use (constraint to exogenous effects, equilibrium procedures etc), an account is given in Gunn (1994). The NMS brings together an extensive network of sub-models representing the interactions of detailed institutional, socio-economic, spatial and accessibility-related factors on final travel demand. The model is essentially an extension of the four-stage model of the 1960s in structure, but disaggregate in estimation and application, and separating out the phases of driving licence ownership, car ownership, tour frequency, mode/destination choice, route and time of day instead of the traditional four stages.

It is important to stress that the studies undertaken during the 1980s *at no time attempted to present the NMS as predicting 'the' future.* A range of uncertainty in key socio-economic and spatial aspects was recognised. For each of a number of feasible levels of these aspects, a set of variants of possible policy measures were put forward.

Gunn and van der Hoorn (1998) set out to comment on *long-run* 'forecasting' accuracy from 1986 to 1996 from two points of view; *firstly*, from the viewpoint of what would have been most likely taken as a 'central forecast' at the time, and *secondly,* having made first-order adjustments to correct for the discrepancies between the central forecasts and the historical development. For the latter, they attempt a *rough correction* of the model output to allow for the differences between the assumed input levels of key variables and the actual outturn. Gunn and van der Hoorn (1998) identified *five* factors for which adjustments should be made. These are:

- *Scale effects:* such as population, households, schoolchildren and students, workforce.
- *Car ownership effects:* we have seen that car-ownership forecasts were too high; reduced ownership must reduce car use, other things being equal.
- *Cost effects:* the differences in fares and car variables costs will have had an impact on modal competition.

- *Travel time effects:* changes in journey speeds by mode will also have an effect on intermodal usage.
- *'System shocks':* in our case, the most identifiable 'shock' was a 'free' student card for public transport.

Comparisons of Adjusted Demand with Outturn

Fig. 2.1 and 2.2 illustrate the comparison of the adjusted model forecasts with actual survey-based outturn. The results (analysed further in Gunn and van der Hoorn, 1998) give indication of a reasonable fit, at least at the level of total kilometers of travel by the major modes and for the major purposes.

The initial conclusions from examining the scenario forecasts were that car-demand was well predicted by the model, but that all other modes were underestimated. Correcting for differences between input assumptions and actual outturn dramatically reverses this conclusion. The non-car-modes turn out to be well-represented by the model (except that exceptional marketing strategies, in this case the student Card, are not considered), and the major deficiency in the system seems to be the failure to predict growth in car-related leisure journeys. In general, the 'priced' modes of public transport and car-driver are somewhat low in the model prediction.

Gunn and van der Hoorn (1998) suggest an explanation of this which relates to the representation of income, or wealth, in the model structure, and in the adjustments that could be made to model coefficients to allow for the different income classes.

The conclusions from this work will add to our understanding of the limitations and possibilities of the model. From the older, SVV2 version, further opportunities to examine the 1996 results in more detail may emerge. At any rate, the year 2000 was also a forecasting year, as was 2010. Exercises such as the one outlined above should help to clarify the picture. From such analyses, we hope eventually to be able to identify and rectify the most important of the deficiencies of the NMS model structure. In the meantime, the results we have obtained give us some confidence that the basic approach is sound; the model behaves sensibly, and in its role as a touchstone for the *comparative performance of policy bundles* (the role for which it was used in the SVV), it seems to produce realistic results.

As an *absolute forecasting tool*, any model is limited by the imagination (and psychic powers) of the modellers. Student Cards, like wars and disasters, are perhaps never to be anticipated. However, it does seem that allowance for changing travel demand with increasing wealth would be a worthwhile extension of the existing model. More detailed analysis, and subsequent analysis in coming years of other forecasts, will doubtless throw up many more ideas; we look forward to reporting these as they arrive.

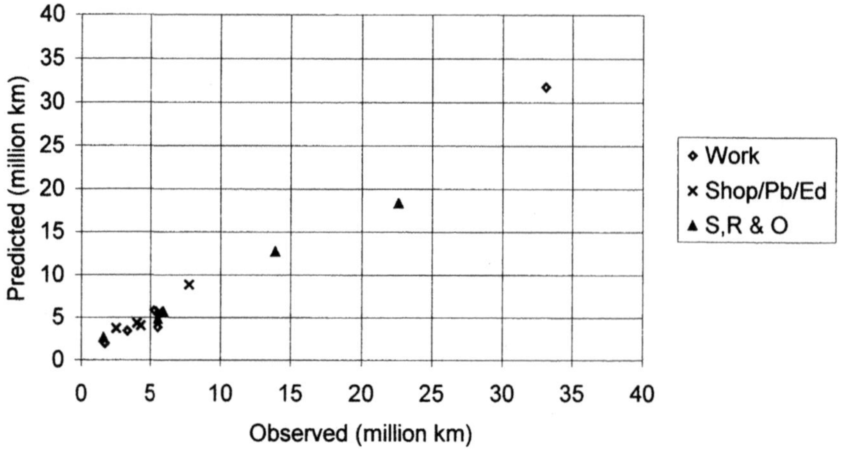

Fig. 2.1. NMS '86-'96 validation for the adjusted model in million kilometres per purpose combination

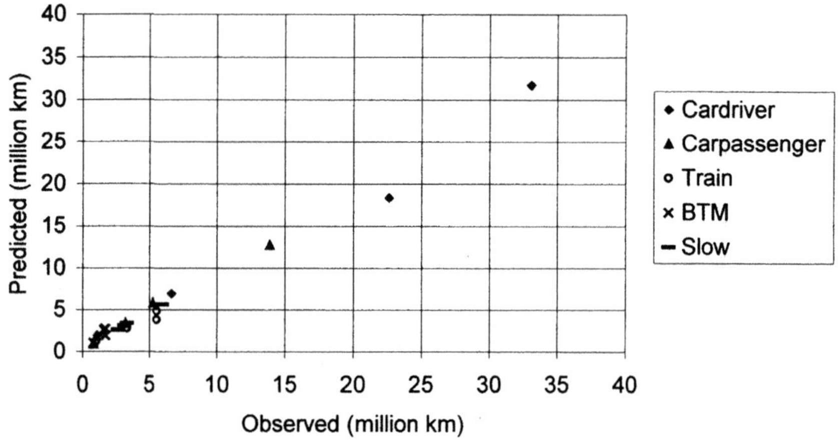

Fig. 2.2. NMS '86-'96 validation for the adjusted model in million kilometres per mode combination

3 Updating and Extending National Models

Andrew Daly
RAND Europe
Newtonweg 1
2333 CP Leiden, The Netherlands

3.1 Introduction: The Problem

The objective of this paper is to discuss the possibilities available for updating and extending the life and capability of national travel demand models. It is based on the experience of the author and of Hague Consulting Group (HCG) in several countries but most particularly in The Netherlands. A final section focuses specifically on the current major project to update and improve the Netherlands National Model (referred to subsequently as NMS).

A national travel demand forecasting system represents a substantial investment in money and time by the national planning authority. While modern techniques of data collection, model development and implementation can help in reducing the costs, it remains inevitable that the modelling of a large and complicated transport system in any detail requires a large and correspondingly complicated model. Once a travel demand model system is in existence, therefore, the client organisation will wish to retain the model in operation to investigate the planning issues that may arise over the subsequent months and years. National Models are rarely decommissioned (Daly, 2000).

Often, the construction of a national model system will be motivated by the needs of a particular planning project; for example, in the Netherlands, the NMS was set up to meet the needs of the Second Transport Structure Plan. Even in these cases, however, the model system can have a substantial value after the completion of the specific project for which it was developed and in appropriate circumstances much more value can be extracted from the initial investment than was obtained by the specific project in question; the NMS has now been in continuous development and use for other studies for 15 years, both before and after the completion of the Structure Plan for which it was developed, and both application and development are expected to continue for some years yet.

In this context, it is natural to consider how to maximise the possibilities for use of a model system after completion of the project for which it was developed. What are the characteristics of a national model that make it re-usable in this way? What can be done to extend the capabilities of a model for future use? Can a model be adapted to future circumstances that are different from those for which it was developed? In the limit, when is it possible to use an existing model and when is it necessary to develop a new one? In summary, to what extent can existing national models be used to evaluate policy different from that envisaged in the model design?

Typically, a national travel demand model is to be developed for a client by a development team to meet an explicit series of policy objectives and a series of applications of the model is then to be made to support decision-making with respect to those objectives. Subsequent to the initial applications, it will not be a great problem to apply the model to investigate policies that vary to a moderate extent from those considered in the model development. However, the applicability of the model for more widely varying objectives may be called into question for several reasons.

- *Policy changes* that the client wishes to investigate but that it had not specified prior to model development can arise for a wide range of reasons: perhaps because of proposals put forward by pressure groups or changes in the balance of financial and environmental objectives. There may even have been a lack of foresight on the part of the client or the development team or it may not have been possible for time or resource reasons to complete the original model development. These are merely examples from a long list of circumstances in which new policy issues can arise beyond the scope of the model that has been developed.

  In addition to policy changes by the client or its critics, there may of course be important changes in the exogenous variables. For example, fuel prices may change or taxation regulations may make certain types of trip (e.g. long distance commuting) more or less attractive.

- *Market changes* may occur, in particular the appearance of new competitors in a market; these may well be brought about because of technological innovations, such as the introduction of fast train services or the appearance of alternative-fuel vehicles in the car market. Alternatively, level-of-service changes may occur or be considered, e.g. price changes because of government regulation, that suggest that application of the model would take it outside its range of applicability.

- *Time passes*, as ever, bringing about changes in the size and nature of the transport market and initiating the suggestion that the model is out of date. Social norms may change, bringing about changes in attitudes with respect to noisy vehicles or those using cleaner fuels.

  Several national models work as 'pivot point' models, that is, predictions are made of *changes* relative to a known base, often expressed in the form of origin-destination trip matrices. An important adaptation of a national model to the passage of time is that these base matrices can be updated; for example, this has recently been done in the case of the Netherlands National Model. However, the procedures for calculating or updating these matrices are complicated and involve no specific innovations of modelling. Accordingly, adaptations of the matrices are not considered in the present paper.

- *Technology advances* within the modelling environment, so that it may be advantageous to adapt the model to take account of improvements in computer hardware or software or to take advantage of advances in modelling knowledge that were not available when the original development took place.
- *Institutional changes* within the client organisation may require the model to adopt a different role, emphasising sketch planning, for example, or to support more graphical presentations of the forecasts made. It may be necessary to transfer the model to a different group of analysts, presenting new requirements for documentation and training.

In any of these cases the modeller may well be faced with the question of the applicability of the (expensive) model system in changed circumstances. The objective of the paper is to describe a range of techniques that the modeller can deploy to maximise the possibility that a model system can be applied with reasonable reliability in changed circumstances.

A key component of the modeller's armoury in such circumstances is the collection of new data and this is considered in Section 3.2. Section 3.3 is concerned with the ways in which policy can be formulated to facilitate model application. In Section 3.4 consideration is given to issues of model design that can influence the success or otherwise of the techniques discussed in Sections 3.2 and 3.3. The final main section gives a brief description of recent developments concerning the Netherlands National Model, which illustrate procedures for a major update. This information is also presented out of general interest. Finally, the main conclusions of the paper are brought together.

3.2 Data Collection or Enrichment

The issue considered in this section is that of a model system that has been developed from a given data base, which may be extended to allow the model to be operated in changed circumstances. Three types of data commonly considered for this purpose are discussed: Stated Response data, aggregate data such as counts and *en route* or intercept surveys. Home interview surveys, e.g. with trip diaries, are less commonly used for this purpose because they are expensive and because they are difficult to focus onto specific types of behaviour; however, when a suitable survey exists that was collected for another purpose, in particular national travel surveys, these can sometimes be exploited for the updating of national model systems. The national travel survey in the Netherlands (OVG) is being used as the basis for the recent updating of the Netherlands National Model, which is described further in Section 3.5 below.

This process can be considered to be that of data enrichment: the original data used to develop the model is enriched with subsequent information designed to focus on issues that have become interesting because of subsequent developments. Only when large new data sets are available as in the Netherlands case, can this be used for the independent development of a replacement model system.

3.2.1 Stated Responses

In this context, the term Stated Response (SR) is used for a range of interview techniques including stated choice, preference rankings and contingent valuations (e.g. 'transfer price'). Many of these techniques are often described as 'Stated Preference' (SP), a term perhaps best reserved for stated choice and ranking data. These techniques have in common that they place an interview respondent in a hypothetical situation, closely related to his or her actual choice situation and ask for preferences between real and/or hypothetical alternative journeys – possibly including the actual journey. A substantial advantage of the approach is that the alternatives presented can be designed to maximise the information obtained. A close relationship between the hypothetical situations put to the respondent and journeys that have been made is essential to maintaining credibility for the respondents and thus obtaining reliable results from an SR exercise.

In contexts where developments are anticipated but have not yet taken place, or are proposed but not yet accepted in the political process, SR is the natural way to investigate likely changes in demand that will or would follow. The new features of the market can be explained as clearly as possible to respondents and their stated responses obtained. However, because SR gives the possibility of obtaining multiple responses from each individual, making it a relatively cheap procedure, it is also often used to investigate existing situations. Among the various SR techniques, SP based on choice methods, usually binary choices, has gained widespread acceptance as being closest to the reality of traveller decision-making.

Applications of SR procedures for model updating and extension can be made to deal with at least four different types of problem.

- The analysis of *new policy* that had not been considered in the initial development of the model and which most likely is not in operation anywhere in the study area. Here the objective of the SR experiments is to attempt to discover the responses of travellers to changes in variables that had previously remained constant or which had not been incorporated in the model at all. For example, road pricing is not incorporated as an explicit policy option in many of the model systems in use for major conurbations. The extent of changes in behaviour resulting from road pricing cannot be determined from actual behaviour when no such system is in operation.

- Another context in which an SR experiment can be useful is when an *aspect of behaviour has to be added* to the model that was not included in its initial specification. For example, an analysis of time-varying road pricing may make it necessary to introduce into the model a description of car drivers switching times to avoid having to pay the road pricing charge.

- The introduction of *new alternatives* into a transport market will naturally require a travel demand model to be adapted to accommodate them. When the new alternatives have not yet appeared in the market, SR may be the only approach to modelling demand for them. Even when the new alternatives are in operation, SR may present a useful means of modelling the demand for them, although in this case it would be preferable also to utilise Revealed Preference (RP) information on the total use of the new alternatives.

- Finally, SR data can be used to *update a model* when it is believed that a number of its parameters, e.g. those determining the trade-off between time and cost, have become out of date. While this process does not change the specification of the model in terms of the alternatives and variables represented, the updating of the parameters can take account of increasing incomes and changes in preferences arising for other reasons. Practical studies of this type are less common.

Thus in a wide range of contexts, the applicability of a model can be substantially enhanced by the exploitation of SR data. The integration of such data into the existing model structure, however, may require a number of adjustments of various kinds to be made to that model and possibly also to the SR data itself.

One approach is to re-estimate the model in its entirety, adding the new SR data to the existing data bases used for estimating the current model. When new alternatives are being modelled, a suitable approach is the three-stage estimation procedure as set out by Daly and Rohr (1998). This approach has been used in several practical studies.

An important consideration in analyses combining SR data with an RP data base is to obtain a correct overall model. It is inevitable that SR data will contain a number of biases relative to RP data. This means that the overall market shares of each alternative in the model must be related to the description given by the most recent RP data. Further, the impact of changes in explanatory variables must be put on a scale that is consistent with the overall scale against which behaviour is measured. Most often, this means that the RP data must be used to define the overall scale of the model as well as the base market shares. Alternatively, aggregate data may be available that can be used along the both SR and RP data to define overall market shares.

A final set of issues arising in the integration of SR data into a forecasting model concern the context in which the extension of the model is to be set. These issues primarily concern the necessary simplification of the choice context for SR interviewing. For example, a typical form of SR interview for obtaining information about a new alternative is to ask respondents to take part in a binary SP experiment in which they compare their current alternative with a new alternative. To use this data in a model in which there are more than two alternatives – some old, some new – requires adaptation of the choice model. Simply stated, the probabilities calculated in a binary experiment are not appropriate for predicting choice among many alternatives.

3.2.2 Aggregate Data

An important and cheap source of data for updating and adjusting models is aggregate data. In particular, count data relating to the changing market shares of the main alternatives is an effective source for keeping limited aspects of the model up to date. Some data of this type can even be collected automatically, e.g. from traffic counters or from ticket sales records. In the latter case, there may well be other information attached to the record, e.g. the ticket type, which can be of further importance in giving information about the segmentation of the market.

The main information content of count data of this type is to indicate the market shares for each alternative and the total market size. Adjusting a model so that it gives overall market shares equal to those observed is usually a matter of adjusting the alternative-specific constants. Any convenient process may be used for the adjustment, often trial and error is as effective as any other. However the adjustment is done, alternative-specific constants that give correct aggregate shares satisfy conditions closely related to the first-order conditions for maximum likelihood, so that the degree of sophistication of the calculation process should not be a concern.

Aggregate observation of the total traffic in the market will obviously be a suitable data source for adjusting a model of growth or generation. Indeed, identifying separately the change in demand due to exogenous growth (e.g. from an increase in average incomes) and that due to endogenous generation (e.g. due to improved levels of service) can present difficulties when several variables change simultaneously. Observation of a series of demand levels over an period of time can help to identify these effects separately and therefore allow the analyst to adjust each model appropriately.

An example of this type of application is the use that has been made of tabular data from the national travel survey for the approximate updating and regional calibration of national and regional models in the Netherlands. The National Model has been updated twice by these methods but is currently being re-estimated completely, as is described in Section 3.5 below, because the approximation inherent in the use of aggregate data was no longer considered acceptable.

3.2.3 En Route Surveys

A third type of data that is very useful in understanding changes in behaviour over the period since the initial development of the model is RP data collected by interviews with travellers. The primary uses of this type of data have been in corridor analysis for the operators of one of the competing systems in a corridor and for national models the application of this type of surveys has principally been to adjust base matrices used in pivot-point procedures. For updating a wide-area model the use of *en route* data is not straightforward because of the problems of representativeness.

3.2.4 Other Approaches

Other approaches to model updating can be considered that involve the collection of no data, or very little data, in the specific market of interest. These approaches chiefly involve model transfers, which can be made on the basis of using a small amount of local data to 'calibrate' a model developed elsewhere, or simply applying, by analogy, model results or observed results from other locations.

Analogy procedures can be very rough-and-ready. An evaluation such as 'a good information system would be worth two minutes of travel time' gives an opportunity to evaluate the impact of information improvements, but this approach depends solely on the quality of the assumption on which it is based.

Analogy and transfer approaches can be useful in special circumstances, particularly when time is short, but usually a more acceptable method, based on more substantial local data collection, can be justified.

An intermediate approach can be considered, in which some components of a model are based on analogy or transfer: for example, the relative impacts of variables in the model (e.g. the 'value of time') can be derived from other areas and the overall model adjusted to take account of the limited local data that may be available. The use of such procedures is usually a response to limited time or study budgets but this approach has proved useful in practical studies.

3.3 Policy Formulation

The issue addressed in this section of the paper is how policy can be formulated for analysis by a model, particularly when the model has not been developed to deal with policy of that type. It is convenient to discuss this issue by considering as an example a specific model system through which many types of policy have been assessed.

The Netherlands National Model, the NMS, was initially used to support decision-making concerning the overall framework of national policy, which was then set out in the Second Transport Structure Plan. Subsequently, more detailed plans were needed for a number of specific aspects of policy, such as for policy implementation by regional authorities, for the railways and for the management of existing transport infrastructure. In particular, the detailed implementation of road user charging systems, such as 'Road Pricing', was of interest. The example of the NMS is interesting because in its relatively long life (15 years of application) a number of different issues have arisen, leading to different approaches to extending the model. It is useful to note that the original objective of the model was to help in determining alignments for the strategic road and railway networks of the country; a considerable number of such studies have been made but these are of limited interest for this paper because they require no specific extensions to the capability of the model.

It is important to be reasonable in attempting to extend the scope of application of a model. There is a point at which further extension is not realistic and it is better for some groups of applications to develop an independent model system. This situation has arisen with respect to the NMS in two applications areas.

First, the application for regional policy, in particular the evaluation of specific projects, has been tackled by setting up a series of regional models for different parts of the Netherlands. The models set up – collectively described as New Regional Models (NRM) – have a common base in their description of traveller behaviour; this base is taken from the national NMS (see Gommers and Pommer, 1995). However, the local inputs of transport networks, zonal data, etc. can be much more detailed than would be possible in the NMS. Moreover, the technical separation of the models also makes possible an institutional separation that fits with the organisation of responsibilities between centre and regions in the Netherlands.

Second, while the NMS has been used for several applications for Dutch Railways (NS) and was extended to meet NS' requirements, ultimately a considerable extension in the level of detail was required to deal with train-specific issues. In particular, the combinations of choice of 'access' and 'egress' modes, the choice of stations and the choice of train service (when several operate in parallel) required a degree of extension that was not reasonable within the NMS. Accordingly, the 'ProMiSe' model was developed (Cohn et al., 1996) to incorporate these features. However, like the regional models, ProMiSe makes extensive use of NMS features, in this case for describing travel modes other than the train; unlike NRM, it uses the same zonal system as NMS, facilitating data exchange.

Apart from these two important examples, it has generally proved possible to extend the NMS to deal with the policy issues that were required. An important example was the assessment of time-dependent road user charging, some variants of which are called 'Road Pricing'. It was intended that the policy would suppress car driver trips and in particular peak hour car driver trips, and would therefore have a positive impact on emissions and congestion, as well as on the transport budget. The transfer of trips from the peak to off-peak helps in reducing congestion, but, because peak-hour travel becomes more attractive for certain classes of traveller (broadly, those with a high 'value of time'), the impact on emissions is less than could be expected. A forecast was therefore required of transfer between time periods.

Time switching behaviour was obtained from an SP survey, conducted in two parts to look at the separate impacts of road pricing and congestion on drivers' choice of time of travel. Analysis of all the SP survey material produced a single model of choice of time of day, as a function of cost and congestion differences. This model was then integrated into the NMS at an appropriate point in the existing model structure. The integration required a number of adjustments to the existing model, e.g. to take account of differing congestion at different times of day and to deal efficiently with the interaction of demand and network models that leads to equilibrium congestion (see Daly et al., 1990). In application the extended model gave useful results, showing that the equilibrated impact of road pricing in reducing peak-period traffic was approximately half what might have been predicted with a naïve approach.

Subsequently, a further SP survey studied the likely take-up of 'passes' which might be offered at a monthly price to allow unlimited use of the tolled highway system. This model was also integrated into the NMS in a similar way (although at a quite different point in the structure) to the time-of-day model.

A quite different approach was adopted for the assessment of a policy of physical restraint on parking. In order to assess the impact of this policy an iterative procedure was adopted in which the first-round numbers of car driver trips arriving in each zone were compared with the available capacity, which in some zones would be quite drastically reduced by the policy. For over-capacity zones the car driver arrivals were then diverted to neighbouring zones (where these were not also affected by capacity limits) or to alternative modes. This procedure was found to give a good assessment of the impact of this important policy instrument, which could change the mode split substantially in city centres although it had little effect in outlying areas.

Another approach was used to assess the impact of 'telematics' (teleworking, teleshopping, etc.). The overall impact of these developments had been assessed in an earlier study and the requirement for NMS was to attribute these overall impacts to changes in mode split, trip lengths etc. This was achieved through calibrating additional terms in the utility functions to achieve the required overall effect, using the model only to distribute the impacts across modes, regions and segments of the population.

Anticipated changes in working practices, such as increased part-time working, whether by reducing hours per day or reducing days per week, were also modelled. Here the approach was to compare people in the base year who had different working practices and apply the different travel patterns of part-time workers etc. more widely in the future. Many of these changes affected the travel frequency for work and other purposes.

A much more detailed study was made to assess the impact of traffic management measures aimed at maximising the use of existing highway capacity (Bakker et al., 1995). Here, a series of measures had been put forward, including ramp metering, flow homogenisation, reserved lanes for specific vehicle groups and various kinds of provision of information to drivers. The aim of the analysis was to assess the impact of these measures on congestion at a national level and to predict changes in traffic flows between classes of roads. There was also an interest in any mode switching effect that the measures might cause. Each of the measures was discussed in detail with traffic engineers familiar with the results of the pilot projects that had been conducted for most of the proposed measures. In each case, the impact was translated into terms that the model could accept, such as changes in capacity or speed on network links. To obtain a full assessment of the impacts, it was necessary to improve the assignment algorithm to give a better interpretation of the time lost in standing traffic, a common feature of the network in peak conditions, and the impact this could have on crossing streams.

Impacts that were easier to incorporate in the model were those concerning fuel price and efficiency. A policy was set out providing for an increase in fuel prices in real terms. The cost impact of this policy on traveller behaviour could be translated directly into the model, which contained a cost variable. However, there is a trend towards the use of more fuel-efficient vehicles and it is expected that this trend will continue, partly under the influence of the policy that has been adopted on fuel pricing. Thus the cost impact of fuel price changes, with its impact on mode split, trip lengths etc., has to be reduced to account for increasing efficiency, although this in turn has a welcome effect on emissions, which are not handled directly in the NMS itself.

Many other policy measures have been evaluated. However, those presented here give an idea of the range of policy and the methods by which it has been interpreted for implementation in the model.

For other model systems, the specific policies to be evaluated will be different in detail. However, the range of possibilities evaluated with the NMS indicates that, with imagination, a model can be adapted to cope with a wide range of possibilities. The applications of the NMS in these ways have generally retained the confidence of the professionals working with the model, both in the development team and in the client organisation, and the adaptations have thus enhanced the value of the model enormously. Other model systems can generally also be adapted in ways that are similar in principle, although different in detail, providing the design of the model does not impose too many restrictions.

3.4 Model Design Considerations

In the previous sections the scope that can exist for extending the model by collecting new data and extending the application of an unchanged model have been illustrated and a number of the techniques that can be employed in those contexts have been described. For the initial design of the model, however, it is interesting and useful to see to what extent the design can be set up to facilitate future flexibility. What is it that makes a model adaptable?

The most important characteristic in the respect is the transparency of the behavioural mechanisms that are represented in the model. If it is clear what the model represents it will be easier to extend it to represent new behaviour, take account of new variables or new alternatives, without impairing the performance of the existing components. An example of the clarity that is needed in this respect would be to base the model on a theory of utility maximisation by individual travellers. Changes in the model can then be determined on the basis of utility theory. For example, to introduce a new alternative a utility function needs to be formulated that can be considered against the utility functions of the existing alternatives; the correlation of the utilities can also be considered to throw light on the appropriate structure of the model. Models based on other clear paradigms could also be extended to deal with new behaviour in similar ways.

Another important aspect of model design that tends towards making the model easily adaptable to future changes is clarity in its breadth of coverage. It should be absolutely clear what is included and what is excluded, and the basis of the assumptions on which these choices have been made.

A further point that is quite important for imaginative application of models is the form in which the various variables appear. Much recent research has noted the improvement in fit that can be obtained from non-linear transformations of the explanatory variables, using Box-Cox transformations, for example. However, such transformations generally tend to impede flexibility in the application of the model. Experience suggests that linear formulations of variables are much more flexible than even simple transformations such as (say) logarithmic or piece-wise linear. Of course, when behaviour cannot reasonably be explained other than by a non-linear transformation then the model will have to contain the transformation, but when the difference is marginal, considerations of applicability should add weight to a choice for simplicity. Linear form makes it much easier to represent changes in the utility of an alternative in a clear and straightforward way.

In this context, as in many other contexts in which statistical estimation is used to set up models, it is important to achieve a reasonable balance between theory and data. Maximising the fit of a model to the data is comparatively easy; achieving a reasonable fit with a model that is behaviourally reasonable is a major challenge. In the judgements that have to be taken, simplicity is always a consideration. The contribution of the present paper to those judgements is to emphasise the importance of simplicity for ensuring transparency and ease of adaptation of the model.

It is often the case that separate alternatives are influenced by variables that are measured in directly comparable units. The obvious example is cost, which applies to most alternatives in most choice models. It is a great help to flexibility in application if the form and coefficient of these variables is consistent across the alternatives. When this form of consistency is present in the model, it becomes possible to implement changes that are consistent across alternatives.

Finally, the variables appearing in the model should as far as possible be those that are truly believed to influence behaviour. The inclusion of variables that are 'strange', perhaps representing some spurious correlation observed in the data, or of proxy variables that take over part of the influence of a variable that is not included in the model, will increase the difficulty of making flexible forecasts of new policy. Only when it is possible to be confident that the model truly describes behaviour is it possible to be confident that interpretations of policy are also correctly implemented.

In conclusion, flexibility in application requires as far as possible a simple model, with the main effects interpreted in standard ways with linear forms. Of course, the accurate representation of behaviour may often suggest that more complicated model forms should be used, but this will inevitably detract from the possibilities for applying the model widely and reliably.

A clear documentation of the model system and a careful and clear explanation of its capabilities to the client organisation will also help in enhancing flexibility in application. If users (and potential users) have a good idea of what a model can do, they will be more inclined to suggest further applications and imaginative enhancements that will permit increased flexibility.

3.5 Updating the Netherlands NMS

From 1997 to 1999 a major project was undertaken to update the central behavioural models incorporated in the Netherlands National Model, the NMS (Bakker et al., 2000). This project has provided new versions of the core models of travel frequency, destination choice and mode choice. Additionally, several of the basic sources of data used by the model have been substantially updated and improved. Since 1999, the new models have been applied in a large number of policy studies.

This project was conducted in parallel with a further major project that updated the 'base matrices' which are used in the forecasting of car traffic by the NMS. The behavioural models of the NMS are used only to predict *changes* in traffic levels, relative to origin-destination matrices set up as much as possible from observed traffic flows and roadside interviews. The present description, however, concentrates on the specifically modelling aspects of the update.

The main data source used for the models was the national travel survey (OVG). In this survey, a large number of households are interviewed each year, making it possible to base the modelling on a single year's data: for the recent study 1995 was chosen. In that year, 68,000 households were interviewed (by telephone), giving information on all of the trips made on a given day by their 168,000 members; a total of 612,000 trip records was collected. To preserve confidentiality, the data supplied to the modelling team by the Central Bureau of Statistics was randomly changed in a small number of variables for a small number of records. These changes did not have a significant impact on the quality of the modelling; detailed tests showed very minor changes in some parameters, but in the key policy variables the differences were very small indeed. The use of a nationally representative survey avoids problems encountered previously with bias or the need to adjust the model to move from one region to another.

Because of the random distortion of the data, the CBS was able to release data that gives much more detail of the locations of trip ends than had been possible previously. Locations are now coded to the level of post codes, approximately 3,000 in the Netherlands, which can easily be aggregated to the 1,308 zones actually used in the modelling. Previous releases had been at the level of municipalities, about 600 in the Netherlands, which had to be split approximately to the model zones. The increased accuracy of location coding greatly enhances the accuracy of the modelling and made it feasible to use the OVG for this type of detailed modelling, rather than requiring a specific survey as had previously been necessary.

The opportunity was also taken to enhance a number of other aspects of the data input to the model estimation.

- The highway network was extended to include a large number of roads that could not previously be incorporated because of software limitations. A total of 36,000 links were included in the network used in the modelling.
- A major new survey was undertaken of the provision and price of car parking space.
- The processing of the rail network was improved, making use of the new TP+ software (a commercial transportation planning package replacing the widely used MinUTP programs), to take better account of parallel services operating between the same stations.
- Data describing bus, tram and metro services was taken from a public information service to obtain a more detailed and accurate description. This data

is used to describe both travel by 'BTM' modes from origin to destination and the use of BTM modes for access to the train.

- New cost functions were calculated for the public transport networks to take account of more detailed information about the travellers, such as their age.
- 'Slow' modes (walk and cycle) were described, as in previous versions of the NMS, by straight-line distance taking account of some physical barriers, mainly water. The opportunity was taken to revise and improve the representation of these barriers.

These data improvements are typical of the opportunities that are given by a major revision of a model system. It would not usually be acceptable to make such changes piecemeal, because of their limited separate value, inconsistency with model estimation and inconsistency with previous forecast results. However, when the model is being changed substantially, a number of changes, which may have been seen as desirable for some time, can be implemented.

The main work that was done was to re-estimate the models describing the choice of mode, destination and travel frequency. These are described by 8 independent models, each dealing with a specific class of travel, e.g. home-work commuting, education trips by children, etc. The modelling of childrens travel could be substantially improved because of extensions to data collection in the OVG survey. Because of the much larger data set than was available for the original estimation of the model, it was possible to introduce many more socio-economic variables into the models: a total of 490 person types was recognised, specifically introducing income variations into the model, compared with the 16 of the previous model. Additionally, it was possible to represent bus and train as separate modes in the models, rather than merging them into a single public transport alternative.

The models are generalisations of the familiar 'logit' form; tests were made of the appropriate structures for these models; the structures vary between the travel purposes. For example, for travel to work, mode choice was found to be best modelled as more elastic than destination choice, while for shopping and 'other' travel, the opposite was found.

For comparison with other methods of updating, estimation was conducted in a series of models each of which departed further from the existing model. This series of tests illustrated the extent to which improvements in the sophistication of the updating improve the quality of the model. For example, the first test is simply to apply the unchanged existing model to the new data; the second is to update as if only aggregate data was available, etc. The final test was to make significant changes to the model specification. Each step in the procedure gave a substantial improvement in the model, leaving the final models with parameters that were plausible and much more securely based on the data than in the previous models.

Implementation of the new models was made simultaneously with a number of other changes to the model system, including the use of the new base matrices mentioned previously, a switch to full 32-bit computer processing, etc. The objective was to avoid making a series of changes that would require separate testing, documentation and explanation.

In the 18 months following the successful implementation of the models, a large number of policy analyses have been made. These have shown that the elasticity values implied by the new model system differ from those of the old, but not by orders of magnitude. Generally, where there are differences, the new values are considered to be more reasonable. Specific policy evaluations have also been made, studying infrastructure and pricing proposals as well as national environmental and transport policy.

Most recently, a further updating project has been undertaken, to improve the model of choice of time of travel. These models, which are chiefly used to predict peak spreading under the influence of pricing policy and/or peak hour congestion, are being updated using SP data and advanced modelling techniques to account for the complex patterns of interaction of choice of different time periods.

3.6 Conclusions

In this paper it has been shown that there are many possibilities to preserve the investment that has been made in setting up a national modelling system. By exploiting these possibilities the analyst can extend the effective life of a model as well as increasing its value substantially. The need for extension of the model can arise from a need to consider new policy, changes in the market or changes in the composition of the travelling population. Of course there are limitations in the extensions that can usefully be made: when much more detail is required in some parts of the model, while other parts can be simplified, an independent model may well be more appropriate.

An important means by which models can be extended is by the collection of new data, which is essentially an enrichment of the original data from which the model was created. Particularly useful types of data in this context are:

- stated responses, which can give insight into many aspects of behaviour, including those not yet possible to observe, but require some care in both analysis and implementation to ensure that proper attention is paid to the respective roles of SR and RP data;
- aggregate data, usually counts, which can be used effectively to ensure that the model predicts the correct total market volume and that the shares of each of the alternatives are correct.

En route surveys can also be considered but are often difficult to use for large-area modelling such as that required for a national model.

The formulation of policy for implementation in the model was considered with reference to examples of implementations of the Netherlands National Model. A range of different methods has been used to interpret the different policies. These methods illustrate the range of methods that can be used to apply models to test the application of policy, when that policy cannot be expressed exactly in terms of the variables explicit in the model.

To design a model that can later be adapted for the application of new policy requires, above all, simplicity in the model design to be maintained. Transparency of the behavioural mechanisms, clarity in the assumptions, the use of linear variables wherever possible, consistency of variables and formulation between alternatives and the avoidance of 'strange' or proxy variables will all help to enhance flexibility in future applications.

The recent update of the Netherlands National Model was also summarised and an ongoing further development was introduced.

References

Bakker, D., Mijjer, P. Daly, A. and Hofman, F. (2000), 'Updating the Netherlands National Model', presented to *European Transport Conference.*

Bakker, D.M., Mijjer, P.H., Daly, A.J. and Hofman F. (1995), 'Prediction and evaluation of the effects of traffic management measures on congestion and vehicle queues', presented to *the 7th World Conference on Transport Research*, Sydney.

Cohn, N.D., Daly, A.J., Rohr, C.L., Oosterwijk, W., van de Star, T. and Dam, A. (1996), 'ProMiSe: Policy-sensitive rail passenger forecasting for the Netherlands railways', presented to *PTRC European Transport Forum.*

Daly, A.J. (2000), 'National Models', in Hensher, D. and Button, K. (eds), *Handbook of Transport Modelling*, Pergamon, pp. 421-432.

Daly, A.J., Gunn, H.F., Hungerink, G.J., Kroes, E.P. and Mijjer, P.H. (1990), 'Peak-period proportions in large-scale modelling', presented to *PTRC Summer Annual Meeting.*

Daly, A.J. and Rohr, C. (1998), 'Forecasting demand for new travel alternatives', in Gärling, T. Laitila, T. and Westin, K. (eds), *Theoretical Foundation for Travel Choice Modelling*, Pergamon, pp. 451-471.

Gommers, M.A. and Pommer, J.F. (1995), 'The Dutch Regional Model System: Applications and development', presented to *the 7th World Conference on Transport Research*, Sydney.

4 National Models: How to Make It Happen. The Case of the Swedish National Model System: SAMPERS

Staffan Widlert
Swedish Institute for Transport and Communications Analysis, SIKA
P.O. Box 17 213
SE-104 62 Stockholm, Sweden

4.1 Introduction

In November 1997, the Swedish Institute for Transport and Communications Analysis (SIKA) issued one call for proposals concerning a new assignment program and one call for proposals concerning new demand models and a user friendly model system. The calls were issued on behalf of a joint group of government agencies. Apart from SIKA the group consists of the Swedish National Road Administration, the Swedish National Rail Administration, the Swedish Transport and Communications Research Board, the Swedish Board of Civil Aviation and the Swedish National Administration of Shipping and Navigation. The group is called "the buyer" in this paper.

The calls were sent out for international competition and resulted in a number of received proposals.

A decision about the new assignment program was taken in January 1998 – the decision was to continue to use EMME/2 as today for individual traffic and for regional public transport, but to use the new facility for deterministic assignment for long-distance public transport.

The decision about the demand models was taken in March 1998 when a contract was signed with Transek as main consultant and Hague Consulting Group and others as sub-contractors. The demand models and the user-friendly system shall be delivered in a preliminary version at the end of January 1999 and with a final delivery in June 1999. The contract also includes a number of options for further development of the system after the final delivery.

The total cost for the demand models is approximately 1.4 million ECU. The travel survey – which serves many purposes but which was specifically designed for model development – costs an additional 1.6 million ECU. In addition to these costs comes the cost for assignment programs and also for coding of networks done by the buyer and different types of supply data, which are provided by the buyer.

The purpose of this paper is to discuss the reasons behind the development work, reasons that touch upon both the planning situation in Sweden and problems with the current models used, to present how the work was organised and also which lessons we learned so far. The purpose of this case-study is to give some insight into why national models are born, what it takes to make them grow and what we believe is necessary to give them a long and healthy life.

4.2 The Reasons behind the New System

4.2.1 A Long Tradition of Strategic Planning

In Sweden, resources for infrastructure investment in different transportation sectors are allocated at the governmental level. In this process, projects in different sectors and in different regions have to be compared to each other. Therefore, there is a need for an evaluation of different projects in a comparable and consistent way, which requires a co-ordinated planning process as well as relevant evaluation tools. Responsibility for other transportation issues than the infrastructure is divided between governmental, regional and local level. In most cases the national transportation policy sets a framework for those decisions. Recently, evaluation of transportation policy measures has been more closely integrated with the evaluation of infrastructure investments. This widening aim has put even greater demands on the evaluation tools.

Strategic infrastructure planning has a long tradition in Sweden. Economic evaluation using cost-benefit analysis started as early as the second half of the 60s for the road sector. Forecasts have been used more and more extensively during the years. Strategic infrastructure plans have traditionally covered a ten-year planning period and they are revised every fourth year.

SIKA started its work during 1994. Among the primary responsibilities are co-ordination of infrastructure planning, forecasts for the development of personal traffic and freight transport and development of methods for forecasting and evaluation in the planning of transport and communications infrastructure.

The National Road Administration has a long tradition in strategic planning and in cost-benefit analysis of investment projects. The Road Administration also prepares the guidelines for the cost-benefit analysis in the road sector.

In 1988 the National Rail Company was divided into one organisation responsible for the operation of trains as a business, the Swedish State Railways, and one new public body responsible for the rail infrastructure, the Swedish National Rail Administration. The rail administration shall evaluate the rail investments according to the same socio-economic principles that are used in the road sector. Cost-benefit analysis based on travel demand forecasting therefore has a strategic role also in the rail sector. In fact, for the rail sector the basic travel demand forecasting plays an even more important role than in the road sector, because of the important mode shifts that occurs as a consequence of investments in the high-speed rail network.

In contrast, investment decisions for air and sea transport are taken from the point of view of business economics.

The four agencies responsible for planning the transport infrastructure, the research funding organisation and SIKA have formed an informal joint planning organisation. This organisation, which is led by SIKA, is responsible for co-ordinating the planning activities in the sector. The experiences in this group from the use of the existing models are one of the major reasons behind the new development work.

In summary we can conclude that cost-benefit calculations as a foundation for the evaluation of projects make heavy demands on the models used. The ongoing trend in Sweden with a more comprehensive planning process, which includes investments, maintenance, traffic safety and environmental issues, reinforces these demands. The integration of transportation policy and investment planning works in the same direction. When the results of the analysis are used to divide money between different measures and different modes of transport – as is currently the case in Sweden – a high degree of confidence in the planning tools is required.

4.2.2 Problems in Models Used Today

Today, the buyer-group uses a set of models for personal travel that includes car-ownership models, a model for long distance trips (over 100 kilometres) and five regional models for shorter trips. This set of models forms a National Model that has been used extensively over the years. A number of problems have been detected during the various applications.

The present models for long distance trips are based on data from 1984/85 and there is a general need to base the models on more current data. The present models do not include any models capturing the importance of the type of train used. When the present models were developed there were only "traditional" trains in Sweden. Now there is an extensive network of high-speed trains, a network, which is gradually being extended. These new trains have better design, higher comfort and they represent a different concept. Since evaluation of new high-speed train projects is one of the most important uses of the demand models, this is a serious problem.

Simple checks show that the long-distance models cannot describe the effect of large changes in a realistic way. Even if travel time were reduced to zero for train, a substantial part of the travellers would still choose air for trips where both alternatives are available. This problem is accentuated by the fact that fairly big changes have been made during recent years when the high-speed trains have been introduced.

According to the long distance models, the dominant effects of the new high-speed trains are a change from car to train and generation of new trips. The relative magnitude of the different possible changes is debated and the transfer between different modes (and especially from air) has often been felt to be unrealistically small and the generation of new trips unrealistically large.

The price elasticity of the long distance models is low compared to information from other sources. One possible source for these problems might be the lack of relevant cost data in the 1984/85 travel survey.

The models also have a very sparse specification in the parts explaining choice of destination. Earlier models included migration variables that seemed to have a large explanatory power. During calibration work it has been observed that the present models do not capture the different roles of different cities (county capitals, cities with large hospitals serving a major region, big university cities etc.).

The models for regional trips use travel data from large cities. These data are also getting old (they were collected more than ten years ago). Most of the models were transferred from other cities – only in Stockholm and Gothenburg are the models based on locally collected data. This has led to problems when adapting the models to more sparsely populated areas in Sweden. Partly, the problems might be explained by the fact that the models lack variables for population structure and economic structure.

The present models are not integrated into a single comprehensive system. Instead, the models are run successively one at a time. User-friendliness was not given a high priority when the different systems were developed. The different parts of the system are not fully documented and the systems do not document the assumptions in the tested scenarios fully enough.

To run the present set of models, with a separate car-ownership model, a long distance model and five regional models, and then to assign the results to the network is an arduous task. As an example, evaluation of projects on the main line between Stockholm and Malmö needs results from five different models (three regional models, the long distance model and a separate model for trips outside Sweden that belongs to another organisation). Even in the cases where only one or two regional models are involved, evaluation is time consuming. Typically, more than a week is needed to run one single forecast for a rail project. This severely limits the number of projects that are possible to evaluate fully.

4.2.3 New Data Are Available

The problems in the present models have been known for a long time. A new travel survey was started in the beginning of 1994 and was originally planned to continue until the end of 1998 (it is now decided that it will be permanent). One of the most important reasons for starting this study was to get data for new demand models. This new data set gives a current and much larger data set, some over-time variation and also better cost data.

In the survey, each interviewee is asked to report all trips done during one day. Different characteristics of the trip and the detailed address of the origin and destination are collected. The interviewees are also asked to report all trips longer than 100 kilometres during a one-month period and trips longer than 300 kilometres during a two-month period. The interview includes questions about the individual and the household. Interviews are evenly distributed on all days over the year.

The present response rate is slightly less than 80 %. The addresses in the survey are coded to zones (and co-ordinates when possible). The zoning system includes approximately 10,000 zones in Sweden. Trips outside Sweden to Denmark, Finland and Norway are coded to county (län, amt, fylking). Trips to other countries are coded to NUTS 2 - level if such a coding system is available. The address coding runs in parallel with the interviews. At present, a total of approximately 70 % of the addresses have been possible to code to zones.

There are also data available from other surveys such as the Fehmarn Belt study, the Öresund study and national travel surveys in Norway, Finland and Denmark.

4.3 The SAMPERS System

Above, we have described the widening application areas for national models in Sweden. This means that the National Model System shall be used for purposes like the following:

- demand effects of new infrastructure and new services, both for single projects and for comprehensive investment programs;
- demand effects of changing incomes, different population structure, changes in trade and industry etc.;
- as a basis for calculation of traffic safety effects;
- as a basis for calculation of environmental effects;
- as a basis for calculation of energy consumption;
- as a basis for calculation of accessibility effects;
- as a basis for calculation of regional effects;
- effects of transportation policy like taxation, tax deductions, road pricing, area licensing, road tolls, rules for company cars etc.

Since one of the most important purposes is evaluating infrastructure investment projects (especially for rail), it is obvious that the models must work on a detailed zonal and network level. The use for policy analysis, on the other hand, demands that relevant policy variables are included in the model and that their impact is captured in a realistic way.

It is not the purpose of this paper to describe the models in any detail. We confine ourselves to a short overview of the main characteristics of the models. The models cover trips within Sweden and to neighbouring countries in detail and trips to and from other parts of Europe in a coarser way. The system covers all trip lengths, both long distance trips and very short local trips. The system includes all non-professional trips and all trip purposes. All the traditional choices are included, plus models for choice of time and class for long distance public transport trips. The system operates on approximately 8,500 zones for local and regional trips and 700 zones for long distance domestic trips. The user will have a choice between "traditional" assignment based on average headway etc. or deterministic "time-table" assignment for long distance public transport trips.

To fit in with the infrastructure planning process in Sweden the new National Model must be possible to use when the next infrastructure plan is prepared. This means that forecasts from the new National Model must be available by June 1st

1999, and that the National Model must be available for project evaluation at the same time. This implies that the new National Model (including both the demand models and the implementation in a user-friendly forecasting system) must be delivered to the buyer in January 1999. The spring of 1999 will be used for final calibration work, systematic testing and validation and also for education of users. This very tight time-schedule has determined much of the design of the development work.

The new National Model will then be further developed before the next infrastructure plan. We have already started to plan for that long-term development work. The next strategic plan will probably be worked out four years later, which means that the long-term development work must be ended well before the end of 2002.

To make a step-wise development possible, it is necessary that the new National Model system is constructed in a modular way. According to the contract, the system will be prepared to include such parts that cannot be included from the beginning.

The demand models are implemented in a forecasting system. The system obviously consists of different parts internally. There are different models for different purposes, models for car ownership and a network assignment program. The whole system will be as well integrated as possible. Ideally, the whole system would be felt as one single part from the user's point of view. Full integration of the network assignment program is difficult at present though.

The forecasting system will be used by different organisations and by different people with very different experience of forecasting and demand modelling. Their frequency of use will also be very different. Therefore, it is important to have a user-friendly interface that simplifies the task for those with little experience and low frequency of use. It is equally important to design an interface that suits experienced users who work frequently with the system.

The need for output from the model system varies between different applications. In some cases fairly detailed output is desired. To facilitate adaptation of the output, a flexible tool for defining the output from a model run is developed. To facilitate the use of the system, a "scenario manipulator" is included in the system.

The National Model will be in use for a long time period and will be developed successively during that time. The need to design the system so that this development is facilitated has been given high priority during the work.

4.4 Organisation of the Development Work

The group of agencies forming the buyer of the system is bound by a contract within the group. The contract regulates the responsibilities within the group and what will happen if some part do not live up to those responsibilities (for example coding networks within the very tight time frame for the project).

The buyer takes an active part in the development work and has therefore formed a steering group for the development work together with the consultant.

Results from the work are reported and approved successively in the steering group. The buyer also follows the estimation work and the development of the software closely by means of special working groups which report to the steering group.

A major task in the development work has been to code networks and to provide different types of supply data. This has been the responsibility of the buyer.

The buyer has the intention of making the system as freely available to other planning organisations as possible. It is also important that the system can be used by different organisations for consulting work. To make this possible, all copyright to the work carried out in the project goes to the buyer. The buyer also have unrestricted ownership to data produced for the project, source code written for the project and other results.

The buyer group will be permanent for future development and future applications. Formal agreements on responsibilities for updating different networks will be made. The group also intends to issue new calls for proposals for the future development work during 1999.

4.5 Lessons Learned

Even though a lot of effort was dedicated to the formal contracts in the project, we can conclude that even more effort would have been worthwhile. Because of the shared responsibilities in the project – where the buyer delivers networks and other data – it is very important to define the borderline between the responsibilities for the different organisations. The shared responsibility creates a lot of practical and formal problems in the process.

We have chosen a process with competitive bidding for the project. It is obvious that such an approach to some extent can come into conflict with the interest to have a long-term commitment from a local consultant. In a small country like Sweden, it is also difficult to ensure competition and to make certain that there is a healthy market.

The work strongly emphasises the importance of building a professional environment. It is both necessary to have experienced consultants in the area, and to have enough experienced users in the different organisations that are supposed to use the new tool. Therefore, the project has involved a substantial educational effort where modelling theory has been taught to participants from both regional and central agencies.

It is also obvious that the success of both development work and future application strongly depends on the interest and commitment of individuals at the agencies concerned.

5 General Modelling Approaches: Top-down or Bottom-up?

Tom E. Worsley
Russell C.E. Harris
Highways Economics and Traffic Appraisal Division
Department of the Environment, Transport and the Regions
Great Minster House, 76 Marsham Street,
London SW1P 4DR, United Kingdom

5.1 Introduction

Most of the models described in this book include detailed transport networks and disaggregated data on travel patterns – mode of travel, journey purpose etc. In some cases the data are also disaggregated in terms of social characteristics – household type, household income etc. Information on the location of the origins and destinations of the trips made by people and goods provides the basic travel demand matrix. These are assigned, by mode, to the networks used for these trips. The approach to building such models starts with data collection, and proceeds through matrix estimation, model calibration, checking, validation and refinement until finally the model is ready for use in testing alternative policies, plans or strategies. Having been extensively validated (and cost a lot of money to develop?) the model commands a high degree of confidence.

The book is provided with only one example of a top-down approach – the use of the National Road Traffic Forecasts (NRTF) for Great Britain, published in 1997. An important difference between the example from Great Britain and the other models was the absence from the British example of a trip matrix and the lack of a geographically distinct network. Not that these are undesirable; the point is that this approach starts with the questions that decision-makers are asking, and develops model components as necessary in order to answer those questions, starting with what can be done using existing data.

5.2 Conditions for Model Development

5.2.1 Bottom-up Models

Bottom-up models are likely to be successful and of value to their users where there is agreement between those responsible for providing inputs to the model and using the model in respect of the following:

- the provision and sharing of data on demand, networks, services etc;
- the behavioural relationships within the model;

- the process for updating the model and funding this and other model development work;
- the model form and questions which it is intended to answer.

Agreement to these principles is more likely to be obtained in a centralised system for decision making, and in one which is based on consensus rather than on competition. The need to reach agreement on the form and the nature of the model is of considerable importance since it is usually very difficult to revise an existing model to analyse questions or perform tasks which it was not set up to do.

Other chapters in this book have considered several of the uses to which bottom-up models have been put. These include developing options for new transport infrastructure, optimising rail passenger service schedules and fare levels and assessing the impacts of transport taxes on the travel behaviour of rural and urban populations.

5.2.2 Top-down Models

The top-down approach, of which the British NRTF provides the example discussed in this book, is perhaps more appropriate in a country where provision of transport infrastructure and services is devolved to a number of competing operators. In these circumstances there is unlikely to be a well established forum for gaining consensus about:

- sharing of data;
- funding of modelling development;
- the validity of the relationships (e.g. rail/road cross elasticities) within the model.

The top-down approach may be more suitable where national Governments have no more than very limited powers to influence local transport policies or to change the provision of specific public transport services, as in Great Britain where public transport is largely provided by the private sector.

The NRTF, as presently set up, represents a very restricted model relating only to road traffic (vehicle kilometres) rather than to trips and other modes. However, there are plans for developing this approach by incorporating data on trips by mode and distance travelled.

5.3 Modelling Approach

5.3.1 Applications: The Top-down Approach

Top-down models may be more flexible and thereby enable their users to analyse policy options which were not identified as being available when the model was initially set up. Because they are, in general, more simple and more transparent than the bottom-up alternative, they can often be adapted to provide a broad understanding of the impacts of new policies without extensive model development work.

The results of such tests are, necessarily, only indicative of the likely impacts of these policies, and there is a trade-off between the reliability or accuracy of the results and the speed with which such new policy analysis can be delivered. For some purposes, speed of delivery is a significant advantage on the grounds that decision-making often takes the form of a fast-moving iterative process. In these circumstances there are advantages in using an analytical approach for providing policy advice rather than leaving Ministers and their advisors to make a guess at the likely impact of a new policy.

5.3.2 Size of Country

Present levels of computing power and capacity are unlikely to present any constraints on the size of country which could be modelled using a bottom-up approach. However, it is notable from this book that the majority of countries for which bottom-up national models had been developed (or were in the course of development) are among the smaller EU countries or those where detailed networks are only required for certain regions in which the majority of the population is located. The Italian national model provides an exception to this rule, although it uses a zone size significantly larger than those in the other models described.

5.3.3 Interaction with Regional Models

Many countries make use of regional and local transport models for a range of purposes including the assessment of infrastructure projects and the modelling of the interaction of regional transport and land use strategies. In some cases, such as in Sweden, the regional models were seen as largely independent of a national model. But there exists the opportunity of building a national model on these regional models if such models are set up on a common and consistent basis.

However, given that such models have been developed for a range of purposes and at different times, and often used different model forms, using them as the basis for a national model is often not possible without a great deal of additional work. In these cases a better solution might be a separate national model which shares only data with the regional and local models, and is otherwise separate from these lower tier models.

The case for a top-down national model is likely to be better than the case for a bottom-up model in a country which already possesses reliable local and regional transport models. These local models could be used to examine detailed spatial issues while the top-down national model would examine wider national policies and strategies. However, countries with a strong modelling tradition – namely those with high quality regional models – are more likely to find support for a bottom-up national model and might use the datasets available from the regional models, and hence construct a national model at a relatively low cost.

5.3.4 Approach to Development: Evolutionary or a New Start

A further distinction worth noting between top-down and bottom-up is the decision process leading to the development of the model. In general the adoption of a bottom-up approach requires a clear commitment on the part of those responsible for the provision of data, funding and use of the model, to commit themselves to a new and distinct modelling project. The top-down approach, as demonstrated by the British NRTF, can be developed on a more evolutionary basis once the basic framework is in place. This evolutionary approach may be desirable where there is less clear commitment from potential model users to a model development programme. This approach often provides potential model users with an approximate estimate of the impacts of strategies or policies early on in the process. But it allows for the possibility of giving a more accurate assessment through incremental development of the model, agreement to which is more likely to be obtained if the model is already giving results which can be used by policy makers.

5.4 Conclusions

In conclusion, there would appear to be no single approach to be recommended in all cases, and the model development process may encompass something of both approaches. The distinction between top-down and bottom-up is itself too simplistic. Models can be disaggregated in respect of household and travel characteristics, while having little spatial detail. Conversely, models with detailed networks and trip matrices might give the policy-maker very little information on the circumstances of the trip-maker and hence be of little value in assessing the impacts of policies on separate social groups. The choice of top-down or bottom-up depends upon the particular circumstances in the country where the model is to be applied and the uses to which it is to be put.

6 Deterministic Time Table Transit Assignment

Isabelle Constantin
Michael Florian
Heinz Spiess
INRO Consultants Inc.
5160 Décarie Boulevard, Suite 610
Montréal, Canada H3X 2H9

6.1 Introduction

The basic timetable based assignment problem is to find an optimal path for a passenger, given an origin node, a destination node, and departure or arrival time information. The deterministic transit assignment does not necessarily minimize the total elapsed travel time, but rather uses weight factors and non-time-based cost elements in determining the optimal path. Therefore, the algorithm tracks both time (to determine the feasibility of a path) *and* cost (to determine the attractiveness of a path).

This paper describes the model and the algorithm for this deterministic assignment model and its implementation in the EMME/2 software package.

6.2 General Problem Definition

The timetable information for each transit line is specified via the following attributes (see Fig. 6.1):

- The *headway* is the interval between successive runs of the line (in minutes).
- The *offset* is the departure time of the first run (in minutes after the beginning of the period).
- The *number of runs* is the number of times the line operates.
- The *segment time* is the travel time on a transit segment.
- The *dwell time* is the time spent at a node serving passengers.

A particular run (number *N)* thus starts at *offset+(N–1)*headway*, and its timetable for stops along the route is determined by the segment times and the dwell times.

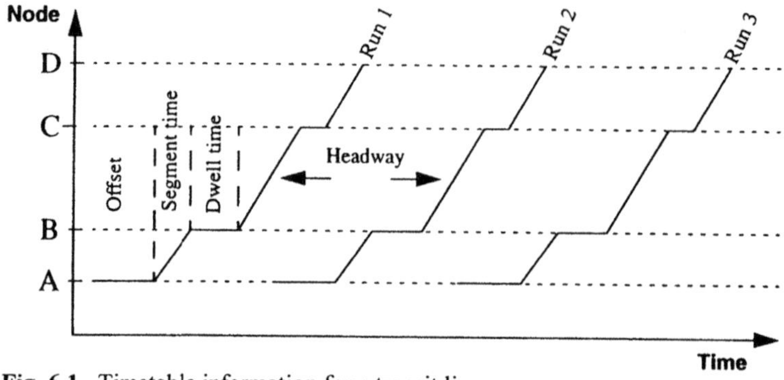

Fig. 6.1. Timetable information for a transit line

6.3 The Deterministic Transit Assignment Algorithm

The deterministic transit assignment problem is solved by using a least cost path algorithm on a space-time network, given by the timetable information plus the transit and auxiliary transit (walk) network. An efficient implementation of such an algorithm must consider the following points:

- As the timetable information is based on repetitions (many runs of the same line), the space-time network can become *very* large (even infinite for continuous services).
- As walking (or using an auxiliary transit mode) is considered to have an infinite frequency, the space-time network theoretically contains an infinite number of corresponding links. Even if, in an optimal path, walking always starts immediately after the departure from the origin or after alighting from a line, the possibility of walking for more than one link creates an exponential number of walk "opportunities" in the space-time network.

The algorithm developed for this problem generates dynamically the part of the network that is actually needed for the computations (instead of explicitly building the whole network, which would be too costly, both in terms of CPU time and memory space). The algorithm computes paths either forward (starting at the origin), for trips with a desired departure time, or backward (starting at the destination) for trips with a desired arrival time.

6.3.1 Description of the Algorithm (for Trips with a Desired Departure Time)

The basic space-time network element on which the assignment is based is an event. An *event* is the combination of a place (node or segment) and a time. A *path* is a chain of events with increasing times. The transition from an event to another event is an *activity*:

From-event	To-event	Activity
node	node	walking
node	segment	waiting and boarding

segment	segment	riding
segment	node	alighting

- At the beginning of the algorithm one or several initial events are created at the origin node, each one corresponding to a possible departure time, based on the granularity and the maximum earliness and lateness. All other events are created as needed during the construction of the least cost path.
- In the usual label-setting manner, each newly created event is inserted into the heap of reached, but not yet scanned events. The heap is sorted according to the minimum cost at which an event can be reached from the origin.
- The algorithm successively retrieves the event with the least cost from the heap and scans it, that is, it checks for all possible next activities (which depend on the event type):

Event type	**Possible next activities**
node	• for all lines stopping at node, wait and board the next vehicle that passes • for all auxiliary transit links leaving node, "walk" to neighboring node
segment	• continue riding on the next segment of the same line • alight at J-node

- For each of the possible next activities, the time t_p and cost c_p of the corresponding potential next event p is computed and compared with the scanned events which already exist at the same element. The algorithm searches for events r which occur at the same element, but earlier than p ($t_r < t_p$), and which satisfy $c_r + (t_p - t_r)*w \leq c_p$, where w is the wait time weight.
- If such an event r does exist, it *dominates* the potential new event p, so that the latter does not even have to be created (see Section 6.4).
- If no event r satisfies the above condition, the new event p is inserted into the heap.

In other words, the above formula tests if the cost of the new event p is less than the cost of an earlier event r plus the cost of waiting from t_r to t_p.

- After a newly created event is inserted into the heap, all *later* events at this element (which are necessarily still in the heap) are checked, using the same formula as above, to see whether they are dominated by the new event. All events dominated by the new event can be removed from the heap and deleted, since they cannot be part of the optimal path.
- The algorithm terminates if an event corresponding to the destination node has been scanned (optimal path found) or if the heap is empty (no path exists). In the case of several destination nodes (which allows the simultaneous assignment of trips from one origin to several destinations, all having the same departure time)

at least one event for each destination has to be scanned for terminating the algorithm.

6.4 Event Dominance

The above mentioned concept of dominance between two events *at the same network element* (node or segment) is used to reduce the number of alternatives to be explored. Since any dominated event does not need to be considered further:

- the algorithm does not include a dominated event (in the heap),
- when including a new element (in the heap), the algorithm discards (from the heap) any element dominated by the new one.

For example, consider the following two events that occur at a given node or segment:

E_1 with time t_1 and cost c_1,

E_2 with time t_2 and cost c_2.

The earlier event E_1 ($t_1 < t_2$) dominates E_2 if $c_1 + (t_2 - t_1)*w \leq c_2$ where w is the wait time weight (set to 1 in the current implementation).

This concept, which is crucial for an efficient implementation of the deterministic transit assignment, is illustrated by the cost/time diagrams in the figures below. These diagrams represent the cost of several events that take place at a given node or segment, and the time at which they occur. The diagonal line starting at an event illustrates the dominance condition, and its slope is w (the wait time weight).

In Fig. 6.2, *A*, *B*, *C*, *D*, and *E* are existing events at a given node or segment. The shaded region shows the area dominated by these events. The algorithm is now considering the two new potential events *X* and *Y*.

- The potential event *X* is in the dominated area, so it does not have to be considered any further. In this case, it is the events *B* and *C* which are dominating *X*:

  $c_B + (t_X - t_B)*w \leq c_X$ and $c_C + (t_X - t_C)*w \leq c_X$.

- The potential event *Y* is in the white region, that is, it is not dominated by any of the existing events. So the event *Y* is created and inserted into the heap. Testing *Y* against the later events *D* and *E* shows that *Y* dominates both of them:

  $c_Y + (t_D - t_Y)*w \leq c_D$ and $c_Y + (t_E - t_Y)*w \leq c_E$,

- so that they can both be discarded.

Fig. 6.3 shows the new cost/time diagram, after event *Y* has been inserted into the heap, and events *D* and *E* have been discarded.

As an example, assume that events *A*, *B*, *C*, *D*, *E*, *X*, and *Y* are different ways to arrive at a given node, by alighting from a transit line or arriving from an auxiliary transit link:

- Event B (for example, arriving with line B) occurs earlier than event X (for example, arriving with line X). Also, the cost of event B plus the cost of waiting until event X (arrival of line X) is less than the cost of event X. So event B dominates event X.
- Event B also occurs earlier than event C (for example, arriving with line C). But the cost of event B plus the cost of waiting until event C (arrival of line C) is larger than the cost of event C. So event B does not dominate event C.

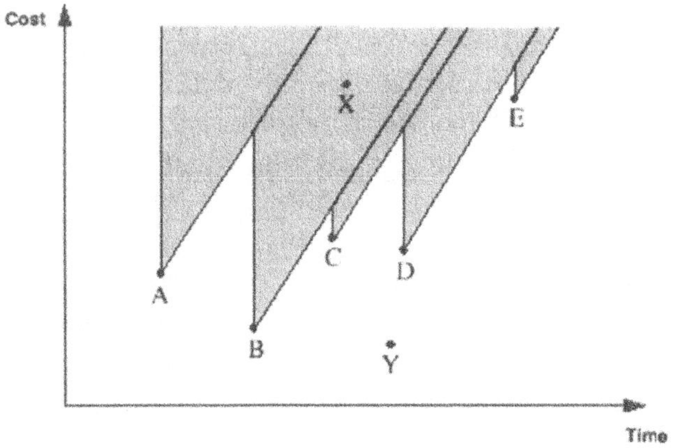

Fig. 6.2. Cost vs. time diagram with potential events X and Y

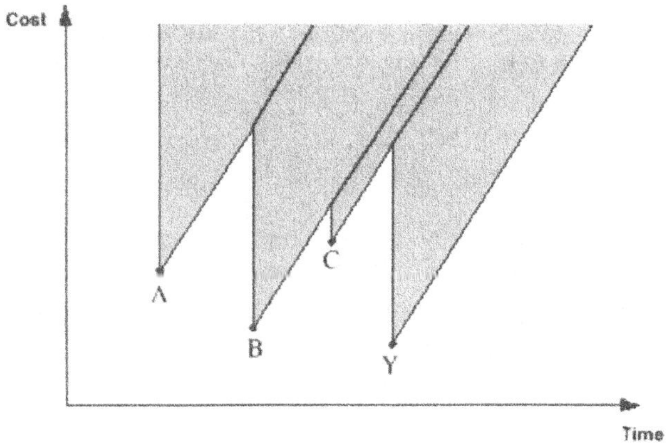

Fig. 6.3. Cost vs. time diagram after insertion of event Y into the heap

6.5 Timetable Information

The timetable information for each transit line is specified via the following attributes:

- The *headway* specifies the interval between successive vehicle runs of the line (in minutes).
- The *offset* defines the departure time of the first vehicle (in minutes after the beginning of the period). The offset is read from **ut1** (transit line user data item 1) by default, but another attribute can be used by setting the appropriate module parameter.
- The *number of runs* specifies the number of times the line operates. The number of runs is read from **ut2** (transit line user data item 2) by default, but another attribute can be used by setting the appropriate module parameter.
- The *segment time* (resulting from the evaluation of the transit time function) gives the travel time on a transit segment. The *dwell time* is the time spent at a node serving passengers. The easiest way to accurately represent a timetable is to code times in a user-defined segment data item, to reflect the time between stops.

A particular run (number *N*) thus starts at *offset+(N–1)*headway*, and its timetable for stops along the route is determined by the segment times and the dwell times.

In the deterministic transit assignment, a transit line definition does not necessarily imply a service operating at a regular headway throughout the entire period (as is the case for the standard transit assignment) but may represent a single vehicle run.

For *continuous operation*, that is lines which run throughout the period, the number of runs is specified as 0. In this case, the line offset specifies the departure time of a run, not necessarily the first.

6.5.1 Example of Timetable Coding

1. *Transit line header coding*. For the following transit line, there is a run every 30 minutes (*headway*=30); the first run leaves 360 minutes (*offset*=**ut1**=360) after the beginning of the period, and there are 6 runs (*runs*=**ut2**=6) during the period.

a ' 1-NW' b 2 30.0 9.06 'aberdeen NW' 360 6 15.033

2. *Transit segment time coding.* Assume that a transit line serves nodes with labels A - B - C - D - E - F, according to the following schedule (the line stops at nodes B, C and E, but the schedule is not given for these stops):

Node	**Run 1**	**Run 2**
A	7h00	7h30
C	7h10	7h40
F	7h40	8h10

If the period starts at midnight (0 minutes), then the line attributes are:

headway:	30	minutes
offset:	420	minutes
number of runs:	2	

6.6 Time Representation

In the standard transit assignment, the time aspect of a trip is simply handled in terms of elapsed minutes. In the deterministic transit assignment, it is also necessary to consider points in time, which can be expressed as minutes after the beginning of the period, or as time of day.

- The *beginning of the period* is specified as the number of minutes after midnight and is used to determine the time of day corresponding to a specific point in time. The beginning of period is a module parameter which has a default value of 0.

Examples for the point in time 330 minutes:

- If the period starts at 0 minutes after midnight, the corresponding time of day is 5h30 (0+330=330 minutes, 330/60=5 hours, remainder 30 minutes).
- If the period starts at 540 minutes after midnight, the corresponding time of day is 14h30 (540+330=870 minutes, 870/60=14 hours, remainder 30 minutes).
- The *time of day* format is *HH***h***MM* where:
  HH hour (0-23)
  MM minutes after the hour (0-59).

For this format:

a preceding – indicates a time in the *previous* day,
a preceding + indicates a time in the *next* day.

Examples: **–23h15 0h17 7h30 8h 14h23 21h04 +1h10**

6.7 Desired Departure or Arrival Time Specification

For each trip, the desired departure (or arrival) time is specified as follows (brackets [ ] indicate optional specifications):

target_time[*–max_earliness*[*$early_pen*]] [*+max_lateness*[*$late_pen*]] [*@granularity*]

- The *target time* can be specified as the time of day (for example **8h**), or as minutes after the beginning of the period (for example **90**):
- The *maximum earliness* (in minutes before the target time) and the *maximum lateness* (in minutes after the target time) determine the range of *allowed* departure (or arrival) times. For example, if the desired departure time is **8h–15+10**, the allowed departure times are between 7h45 and 8h10. The default values for maximum earliness and maximum lateness are module parameters (default 0 minutes).
- The *granularity* (in minutes) determines the spacing of the allowed departure (or arrival) times, with respect to the target time. For example, if the desired departure time is specified as **8h–15+10@5**, the granularity is 5 minutes and the allowed departure times are: 7h45, 7h50, 7h55, 8h00, 8h05 and 8h10. The default value for the granularity is a module parameter (default 1 minute). Note that using a small granularity will result in more precise early/late component of the trip cost (see Section 6.9) but will also significantly increase the assignment execution time.
- The *earliness penalty factor* (per minute earliness) and the *lateness penalty factor* (per minute lateness) are used to compute the early/late component of the trip cost. See Section 6.9. Default values for these two penalty factors are module parameters (default 0).

6.8 Examples of Desired Departure and Arrival Times

The following examples assume the default setting of the assignment parameters for maximum earliness and lateness (0 minutes), penalty factor for explicit earliness and lateness (0/minute) and granularity (1 minute).

6.8.1 Desired Departure Times

13h30	Target departure time is 13h30.
13h30–15+30	Target time is 13h30, but it is possible to leave up to 15 minutes earlier or 30 minutes later (allowed times are between 13h15 and 14h00).
13h30–15+30@5	Same as above, but with a 5 minute granularity (allowed departure times are 13h15, 13h20, 13h25, ...,13h55, 14h00).
13h30–15$1.2+30$.3@5	Same as above, but with penalty factors of 1.2

	per minute earliness, and 0.3 per minute lateness.
14h00–45$1.2@5	Target time is 14h00, but it is possible to leave up to 45 minutes earlier with a penalty factor of 1.2 per minute earliness. The allowed times are the same as above, but the perception of earliness and lateness is different, yielding different results.
–23h15	Target departure time is 23h15 the previous day.

6.8.2 Desired Arrival Times

8h45–20$.4+10$1.2@10	Target arrival time is 8h45; allowed times are 8h25, 8h35, 8h45, 8h55. Penalty factors are 0.4 per minute earliness, and 1.2 per minute lateness.
8h25+30$1.2@10	Target time is 8h25, but it is possible to arrive up to 30 minutes later with a penalty factor of 1.2 per minute lateness. The allowed times are the same as above, but the perception of earliness and lateness is different, yielding different results.
+2h00–60+60@15	Target arrival time is 2h00 the next morning, with a maximum earliness and lateness of 1 hour and a granularity of 15 minutes.

6.9 Earliness and Lateness

Based on the desired time for departure (or arrival), the module determines the feasible space-time path that has the least total cost. For a given path, the time at which the trip actually starts (or finishes) is called the *actual* departure (or arrival) time. Earliness or lateness occurs when the actual time differs from the target time. The difference between the *target* and the *actual* time is defined as *early* or *late time*. The corresponding early or late cost is calculated by applying the appropriate penalty factors. An earliness (or lateness) penalty factor of 0 reflects the traveller's indifference to leaving or arriving earlier (or later) than his target time.

6.9.1 Early and Late Departures

If the target time is the only allowed departure time, the actual departure time may be later (but not earlier) than the target time. This "waiting" at the origin is considered as *implicit lateness*, which carries the same weight as waiting (=1).

If several departure times are allowed, the best one (not later than the actual departure time) is *selected.* If none of the allowed times corresponds to the actual departure time, the time selected is the allowed time which immediately precedes the actual departure time.

For example, if the desired departure time is **8h+15@5**, the allowed departure times (marked • in the diagram) are 8h00, 8h05, 8h10 and 8h15; for an actual departure time of 8h14, the selected departure time would be 8h10; for an actual departure time of 8h23, which is later than any allowed departure time, the selected departure time would be 8h15.

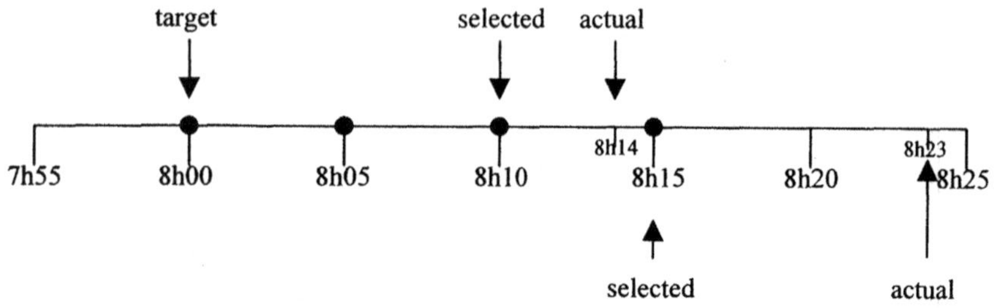

The difference between the *selected* and the *actual* time is considered as *implicit lateness* and carries the same weight as waiting (=1). This reflects the frustration associated with waiting as a result of being forced (by the granularity or the maximum lateness) to leave before the actual departure time.

The difference between the *target* and the *selected* time is considered as *explicit earliness or lateness*, for which explicit penalty factors are specified by the user. These penalty factors reflect the willingness of the traveller to leave before or after the target time, and apply to earliness or lateness within the limits set by the maximum earliness or lateness. However, for a late departure, any explicit lateness penalty factor is only applied if it is less than 1. Otherwise, the wait time weight (=1) is applied, since it is then cheaper to "simply wait".

6.9.2 Early and Late Arrivals

If the target time is the only allowed arrival time, the actual arrival time may be earlier (but not later) than the target time. This "waiting" at the destination is considered as *implicit earliness*, which carries the same weight as waiting (=1).

If several arrival times are allowed, the best one (not earlier than the actual arrival time) is *selected.* If none of the allowed times corresponds to the actual arrival time, the time selected is the allowed time which immediately follows the actual arrival time.

For example, if the desired arrival time is **10h30+30@15**, the allowed arrival times (marked • in the diagram) are 10h30, 10h45 and 11h00; for an actual arrival time of 10h33, the selected arrival time would be 10h45; for an actual arrival time of 10h14, that is earlier than any allowed arrival time, the selected arrival time would be 10h30.

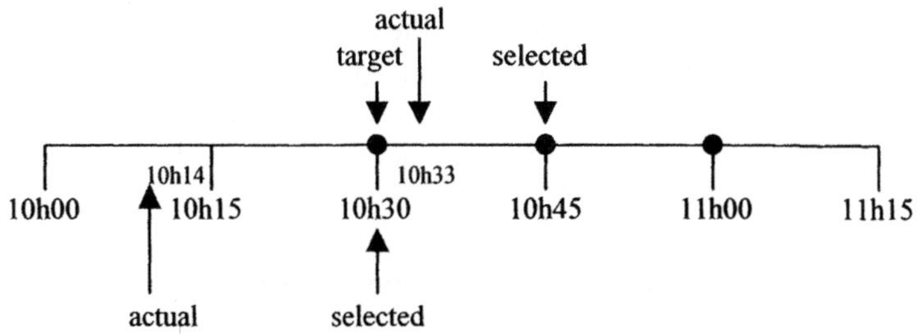

The difference between the *selected* and the *actual* time is considered as *implicit earliness*, and carries the same weight as waiting (=1). This reflects the frustration associated with waiting as a result of being forced (by the granularity or the maximum earliness) to arrive after the actual arrival time.

The difference between the *target* and the *selected* time is considered as *explicit earliness or lateness*, for which explicit penalty factors are specified by the user. These penalty factors reflect the willingness of the traveller to arrive before or after the target time, and apply to earliness or lateness within the limits set by the maximum earliness or lateness. However, for an early arrival, any explicit earliness penalty factor is only applied if it is less than 1. Otherwise, the wait time weight (=1) is applied, since it is then cheaper to "simply wait".

6.10 Examples of Earliness and Lateness

6.10.1 Early Departure (Only Possible if Maximum Earliness Is Not 0)

Desired departure time: **8h–15+10@5**
Actual departure time: **7h53**
Allowed departure times: 7h45, **7h50**, 7h55, 8h00, 8h05, 8h10

Because the granularity is 5 minutes, departure can take place at: 7h45 (not optimal), 7h50 (optimal), 7h55 (too late)... The selected departure time (7h50) is –10 minutes from the target time (8h00). Also, since the actual departure occurs at 7h53, the traveller has to wait (implicit lateness) 3 minutes between his selected time and his actual departure time. The resulting early time is thus –7 (= –10+3).

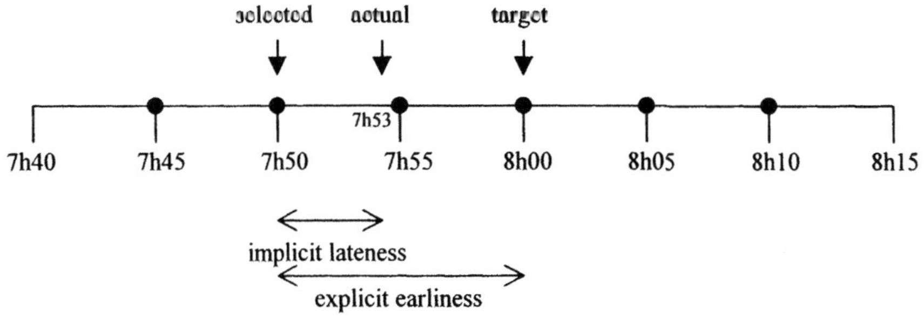

The explicit earliness penalty factor applies to the 10-minute earliness, but a weight of 1 applies implicitly to the 3-minute implicit lateness. With an earliness penalty factor of 1.5, the resulting early cost would be 18 (=1.5*10+1*3).

6.10.2 Late Departure (Maximum Lateness Is 0, the Default)

Desired departure time: **8h**
Actual departure time: **8h09**

The actual departure time is 8h09, that is 9 minutes later than the target time (8h00). This is an implicit lateness of 9 minutes, to which a weight of 1 applies. The resulting late cost is thus 9.

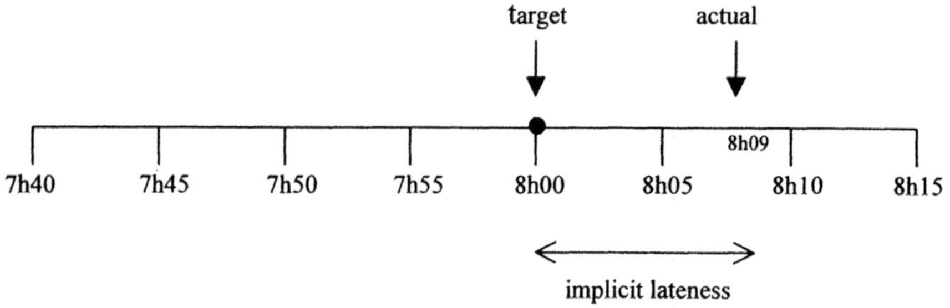

6.10.3 Late Departure (Maximum Lateness Is Not 0)

Desired departure time: **8h–15+10@5**
Actual departure time: **8h09**
Allowed departure times: 7h45, 7h50, 7h55, 8h00, **8h05**, 8h10

Departures can be at ... 8h00 (not optimal), 8h05 (optimal), 8h10 (too late). The selected departure time is 8h05, that is 5 minutes later than the target time (8h00). Then there is an additional wait (implicit lateness) of 4 minutes. The resulting late time is thus 9 (=5+4).

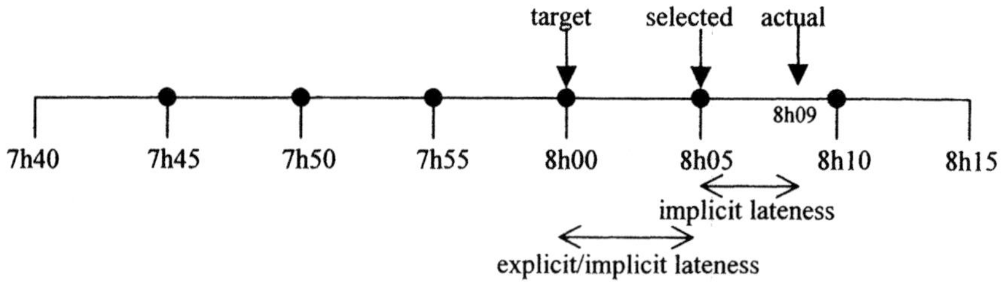

The time between 8h00 and 8h05 can be perceived either as explicit lateness (to which the lateness penalty factor applies) or as implicit lateness (to which a weight of 1 applies).

– If the lateness penalty factor is less than 1 (for example 0.25), it applies to the 5-minute lateness, but a weight of 1 applies to the 4-minute implicit lateness. The resulting late cost is thus 5.25 (=0.25*5+1*4).

– If the lateness penalty factor is greater than 1 (for example 1.1), it is better to "simply wait". The resulting late cost is thus 9 (=1*5+1*4).

6.10.4 Late Arrival (Only Possible if Maximum Lateness Is Not 0)

Desired arrival time: **18h–30+30@15**
Actual arrival time: **18h10**
Allowed arrival times: 17h30, 17h45, 18h00, **18h15**, 18h30

With a granularity of 15 minutes, the allowed arrival times are: ... 18h00 (too early), 18h15 (optimal), 18h30 (not optimal). The selected arrival time is thus 18h15, that is 15 minutes later than the target time (18h00). The traveller is considered to "wait" (implicit earliness) –5 minutes between his actual (18h10) and his selected (18h15) arrival time. The resulting late time is thus 10 (=15–5).

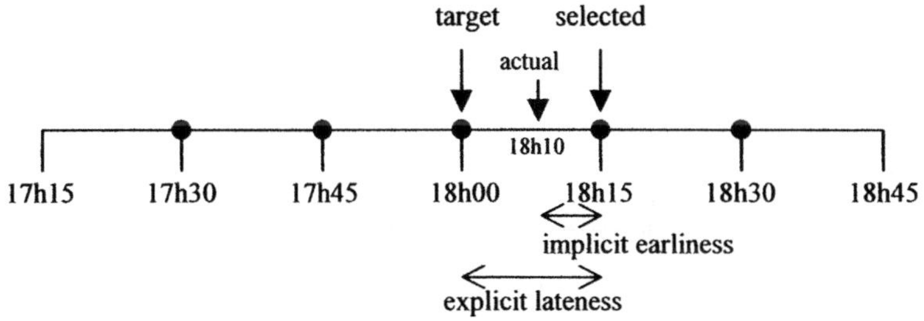

The explicit lateness penalty factor applies to the 15-minute explicit lateness, but a weight of 1 applies to the 5-minute implicit earliness. With a lateness penalty factor of 0.25, the resulting late cost would be 8.75 (=0.25*15+1*5).

6.10.5 Early Arrival (Maximum Earliness Is 0, the Default)

Desired arrival time: **18h**
Actual arrival time: **17h38**

The actual arrival time is 17h38, that is 22 minutes earlier than the target time (18h00). This is an implicit earliness of 22 minutes, to which a weight of 1 applies. The resulting early cost is 22.

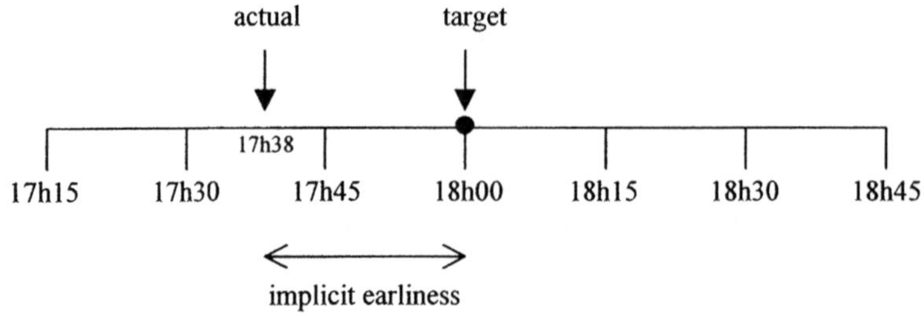

6.10.6 Early Arrival (Maximum Earliness Is Not 0)

Desired arrival time: **18h–30+30@15**
Actual arrival time: **17h38**
Allowed arrival times: 17h30, **17h45**, 18h00, 18h15, 18h30

Allowed arrival times are 17h30 (too early), 17h45 (optimal), 18h00 (not optimal)... The selected arrival time (17h45) is 15 minutes earlier than the target time (18h00). The traveller also has to wait (implicit earliness) between the actual arrival time (17h38) and the selected time (17h45). The resulting early time is thus –22 (= –15–7).

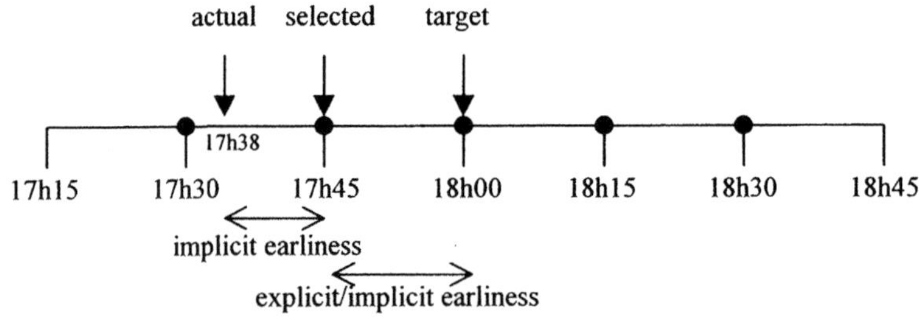

The time between 17h45 and 18h00 can be perceived either as explicit earliness (to which the earliness penalty factor applies) or as implicit earliness (to which a weight of 1 applies).

- If the explicit earliness penalty factor is less than 1 (for example 0.75), it applies to the 15-minute earliness but a weight of 1 applies to the 7-minute "wait". The resulting late cost is thus 18.25 (=0.75*15+1*7).
- If the explicit earliness penalty factor is greater than 1 (for example 1.1), it is better to "simply wait". The resulting late cost is thus 22 (=1*15+1*7).

6.11 Timed Trip Information

The following information defines a timed trip:

- origin node,
- destination node,
- desired time for departure (or arrival),
- volume, that is the number of trips with the same specifications (default is a module parameter).

The trips can be specified interactively, or as a demand matrix.

6.12 Minimum Waiting Time

The *minimum waiting time* is used in determining the feasibility of a path: the arrival time at each boarding node must allow for this minimum time. For example, if the minimum waiting time is 2 minutes, the traveller must be at the stop *at least* 2 minutes before the line leaves the stop. The minimum waiting time must be between 0 and 999.99.

The minimum waiting time may be the same value for the entire network, be node specific (stop dependent), or line specific (line dependent). It may also be node *and* line specific, in which case, the *maximum* of the two values is used (for example, if the minimum waiting time is 3 minutes for node *N* and 5 minutes for line *L*, the traveller who wishes to take line *L* at node *N* must be there at least 5 minutes in advance).

Node/line specific values can be contained in user data items or extra attributes. Depending on the source selected, the user is then prompted for the minimum waiting time value, or for the relevant attribute(s) to be used.

6.13 Boarding Penalties

The *boarding penalty* is used to compute the cost of a path. The penalty applies to every boarding, be it an initial or a transfer boarding. It must be between 0 and 999.99.

The boarding penalty may be the same value for the entire network. It may also be node and/or line specific, in which case the values are contained in user data items or extra attributes. If node *and* line specific is selected, the *sum* of the two values (for line *L* at node *N*) is used.

6.14 Weights for Time Components

Weights can be specified for the auxiliary transit time, the boarding penalty and the in-vehicle time:

- The weight for the boarding penalty must be between 0.00 and 999.99.

- The weight for the in-vehicle and auxiliary transit time must be between 1.00 and 999.99 (that is, they cannot be less than 1).
- The weight for the waiting time is assumed to be 1, and cannot be modified by the user.

The above conditions ensure that travelling uselessly in a loop is never "cheaper" than waiting.

6.15 Demand and Result Matrices

The *demand matrix* to be assigned must be specified. The corresponding departure (or arrival) time is specified later in the dialog.

Each of the following results can be saved in the form of a matrix containing, for each O-D pair processed:

- individual *travel time components*:
- early or late time,
- auxiliary transit time,
- waiting time,
- in-vehicle time,
- *total travel time*, that is, the sum of the auxiliary transit time, waiting time and in-vehicle time,
- individual *transit cost components*:
- early or late cost (see Section 6.9),
- auxiliary transit cost (weighted auxiliary transit time),
- waiting/boarding cost (sum of the waiting time plus the weighted boarding penalties),
- in-vehicle cost (weighted in-vehicle time),
- *total transit cost*, that is, the sum of the transit cost components listed above,
- *number of boardings,*
- *distance travelled.*

6.16 Timetables

The timetables are displayed on a space-time diagram. Space is represented as a sequence of nodes displayed on the Y-axis. Time is represented on the X-axis, and can be displayed as the time of day or as the number of minutes since the beginning of the period. A grid makes the points on the plot easier to identify. Each run is represented by a line that traverses the grid.

The timetable of a transit line is computed using the following line and segment attributes:

- line offset,
- line headway,
- number of runs,
- travel times on segments,
- dwell time at stops,
- layover times, which are considered as additional dwell times at the layover nodes.

6.16.1 Plot Transit Timetable

This option is used to specify the timetable(s) to be plotted (for which transit lines, node sequence, time period, etc.). Examples are given in Fig. 6.4 and 6.5.

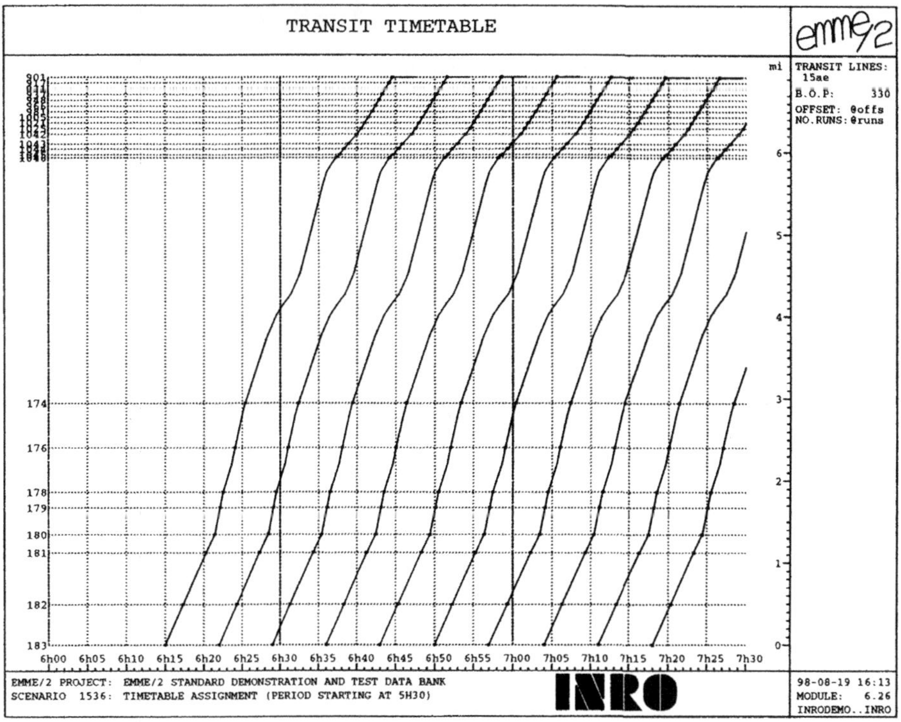

Fig. 6.4. Timetable for line 15ae from 6h to 7h30

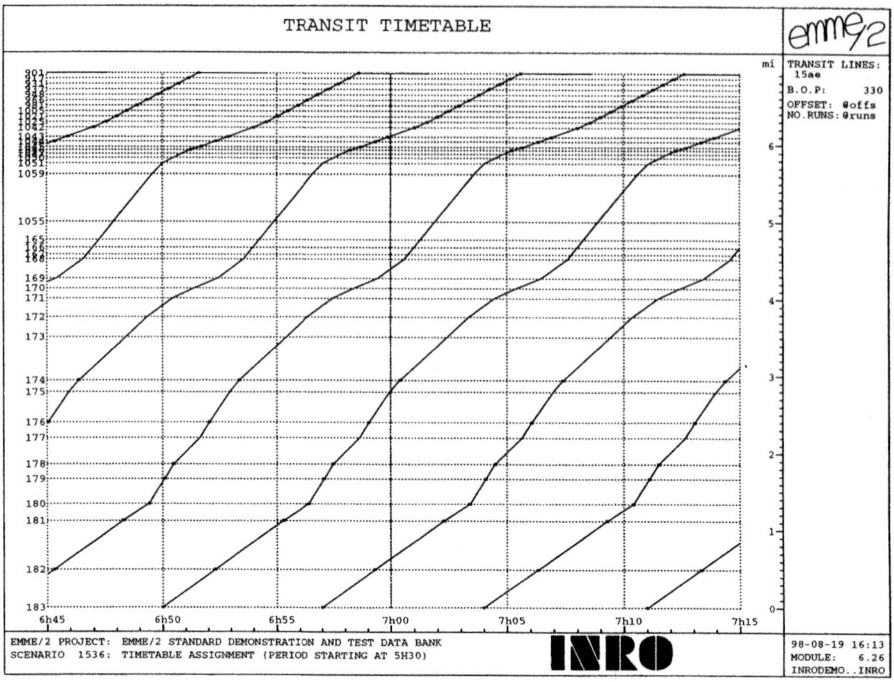

Fig 6.5. Result after changing the timetable period to 6h45 to 7h15

Acknowledgments

This development was carried out by Heinz Spiess. The text is based on sections of the *EMME/2 User's Manual*, Release 9, which was written by Isabelle Constantin, Linda James-Lefebvre and Gabriela Simonka.

Reference

INRO Consultants (1998), *EMME/2 User's Manual,* Release 9.

Part II

National Transport Models

7 Application Areas for the Dutch National Model

Frank Hofman
Ministry of Transport, Public Works and Water Management
Directorate-General of Public Works and Water Management
Transport Research Centre (AVV), Strategic Studies Division
Rotterdam, The Netherlands

7.1 Introduction

Since 1986 extensive use has been made of the Dutch National Model System of Traffic and Transport, abbreviated NMS, in policy evaluation. Although the NMS certainly is not the only available model in the Netherlands it is the most influential model for use in evaluating policy options and preparing policy documents at the national level.

One of the reasons for this success is the need for continuity: the NMS was used during the preparation of the Second Netherlands Transport Structure Plan (TSP-2), which has been the major strategic policy document of the past decade. Several quantitative targets mentioned in the TSP are based on effect evaluations using the NMS. The estimated effects of new policy alternatives should be comparable with past calculations. The use of other models would cause interpretation problems: are differences caused by differing policies or by other model methodologies?

Of course the other reason is the quality and the flexibility of the model. The NMS has proven to give credible results and was several times extended to deal with new policy options (Gunn et al., 1993).

In this paper an overview will be given of the sort of applications the NMS is used for. After a brief introduction to the major characteristics of the model we will classify the applications of the NMS in different types and give some recent examples. Recent applications show a shift in emphasis from environmental concerns to assessing accessibility consequences of policy options. This raises the question of the strong and weak points of the NMS and consequences for future developments.

For an overview of the role of models in general in transportation planning in the Netherlands see e.g. van der Hoorn et al. (1995) and van der Waard (1997).

7.2 A Brief Introduction to the NMS

The main purposes of the NMS are to provide insight into changes in the mobility patterns of the Dutch population and to predict traffic flows on the Dutch trunk road and railway networks.

Application of the NMS takes place within the context of spatial, socio-economic and demographic scenarios and within the context of specific transport policy scenarios (e.g. increased fuel taxes, vehicle purchase taxes, improvement of public transport etc.). Within such scenarios, the NMS provides forecasts that relate to:

- National and regional levels;
- Car traffic, slow modes (pedestrians, cyclists and mopeds) and public transport;
- Average working day, peak and off-peak periods.

The NMS is responsive to changes in:

- Land-use factors;
- Accessibility characteristics;
- Socio-economic factors;
- Demographic factors.

The NMS provides forecasts for the long term, a typical period is 20 years. It can produce forecasts for different travel purposes and for specific groups of people and households.

The NMS consists of a series of sub-modules, each of which is based on model structures compatible with rational individual (and household) decision making, and which are calibrated using disaggregate data on individuals and households.

The model uses a pivot point approach, in which growth factors are applied to base matrices for the car driver mode and the train passenger mode. The choice models are used to calculate various probabilities for all the individuals and households in the so called prototypical sample. This is a sample, drawn in such a way that there are sufficient observations of a certain number of household categories to make statistically reliable statements about the travel behaviour of that category. For every zone expansion factors of the strata of the sample for these household categories are calculated.

The model system consists of a set of connected choice models.

The first models applied are the driving license and car ownership models. The results of these models, probabilities for car ownership/license holding status of a household and probabilities for license holding of the individuals are adjusted to national totals obtained from so called cohort based models (van den Broecke, 1988) or, for car ownership, a model, FACTS, that attempts to predict the composition of the car fleet in terms of weight classes, fuel types etc. and is aimed at forecasting the pollution caused by car traffic (Pronk and van Wee, 1992).

Consecutively tour frequency models are applied. These compute the number of tours for every purpose and depend on the household structure, the license/car ownership status of the household, the license holding of the individual and the educational level of the individual.

The mode and destination choice models (NSES) predict the distribution of tours over possible combinations of destinations and modes. This model depends on the accessibility for every mode and on attraction variables that are purpose specific. The geographical detail of the NMS is greater in this sub-module: 1,302 sub-zones are used instead of 345 zones.

Using growth factors resulting from applying NSES for both the base year and the forecast year in combination with the base matrices for the car driver mode and the train passenger mode, assignment matrices are obtained. Cars are assigned to the network using a capacity restrained assignment technique that takes account of both a possibly restricted inflow on a link due to congestion and a restricted outflow capacity due to congestion on downstream links (Bakker et al., 1994).

Using a time of day model changes in departure times due to congestion or period specific costs (road pricing) are calculated for the car driver mode.

Train passengers are assigned to the trains using the Prolop model which is a special purpose software tool that is owned and operated by the Dutch Railways.

Using different feedback loops, the model is designed to search for an equilibrium between time of day choice and congestion and period dependent costs and between total demand and congestion and cost of travel.

An overview of the structure of the NMS with the emphasis on its further development in response to changing needs can be found in Gunn et al. (1993). A more detailed account of the structure of the NMS can be found in Rijkswaterstaat (1992). Because the model was improved on several occasions since the publication of this report, the information given is not up to date anymore.

7.3 Strengths and Weaknesses of the NMS

In the NMS the zoning system consists of 345 zones, a dense system in the highly urbanised Randstad area and larger zones in the northern, eastern and southern parts of the country. The level of detail of the road network is in accord with this zonal system, comprising the trunk road network, the main other inter-local connections and main roads in the urban areas.

Although the network shows a lot of detail, the NMS is not considered a suitable tool for regional infrastructure assessment studies.

The NMS is not an activity-based model. Relations between various activities are not captured and restrictions caused by space-time relationships are not properly accounted for. *This means that for instance the effect of flexitime arrangements can only be estimated by making assumptions about the consequences on in this case the number of commuting trips.*

Also the effects of changing the institutional context on activity schedules and thus on the timing of trips cannot be evaluated using a model like the NMS. Currently, a special model is under development to provide insight into these issues (see e.g. Arentze et al., 1998).

In the NMS feedback of congestion to the mode and destination choice models might suppress demand, or in the case of measures that reduce congestion, might increase demand in the peak hours on the trunk road network. These phenomena reflect the so called latent demand. Empirical evidence with respect to the size of latent demand is hard to obtain. Actually, this part of the model was to some extent based on expert judgement.

In my view there is a large amount of uncertainty in the estimated size of the latent demand effects.

The NMS is estimated on cross section data. *It does not capture the dynamics of individual behaviour*. For instance, habit or anticipation are not included in the models.

The effects of a measure according to the NMS might include changes in route choice, time of day choice, choice of mode or change of destination. How much time should elapse before these change in behaviour actually take place is not a result of the calculations.

In the model the land use influences transport, but land use is not influenced by accessibility. Regional economic development is part of the scenario input to the model.

The NMS is sensitive to a wide range of influencing factors of transport. It can not only deal realistically with future trends in prices, income, population, the number of households, but also with changes in the composition of households in terms of age/sex combinations and occupation patterns.

7.4 Application Areas

7.4.1 Assessment of Rail Options

Nowadays there is a separate Task Organisation of the Ministry of Transport, Public Works and Water Management, responsible for capacity management, capacity planning and railway safety, called Railned. The Transport Research Centre (AVV) co-operates with Railned to evaluate infrastructure options for the railway network. The NMS is used to obtain forecasts for the rail demand.

The socio-economic environment for these studies is based on scenarios that are available at that time and are used for policy evaluation purposes. Assumptions about the implementation of policy measures relevant for calculations of train demand are based on ideas about current policy plans.

The procedure is that Railned makes assumptions about the quality and price of the train services that will make use of the infrastructure. These assumptions are input to the model and growth factors for every purpose and origin destination combination are calculated and applied to the base matrices for train passengers that are owned by the Dutch Railways. The resulting trips are assigned to the train services. Although there is no time of day model for public transport travellers it is possible to take account of differing frequencies of trains during the peak and off-peak by creating different images of the day: *peak* (both legs of the tour in the peak), *half-peak* (one leg in the peak and one leg off-peak) and *off-peak* (both legs off-peak).

The NMS output is aggregated to mode specific information with respect to kilometres and tours. The assignment results that are presented are based on the special purpose assignment model Prolop.

In the ideal case the resulting demand is confronted with the assumed quality of the services which should be diminished in case of a too small demand.

7.4.2 Evaluation of Policy Scenarios

In this case the effects of a package of policy measures are calculated within the context of one or more scenarios describing the socio-economic environment. During the scenario development these scenarios are usually translated to the detailed level of 345 NMS-zones, which is the required level of detail of input for the NMS applications.

The next step in the application of the NMS is the translation of policy measures in input of the NMS. In some cases, such as raising the fuel prices, this is easy. In other cases, such as the introduction of motorway signalling, this is not a trivial exercise. In this case, the effect of motorway signalling on capacity that was obtained in evaluation studies, is used to represent the measure.

The NMS generates an enormous amount of output data. The data that are presented to the policy makers are usually related to network use, congestion, or total mobility by mode/purpose combination. These data are aggregated to the level of the Randstad area, or the total country and are in case of network related information, sometimes presented by type of road and by period of day.

As mentioned before, the NMS is not the right instrument to use for regional infrastructure assessment studies. Nevertheless, the NMS is used for evaluation of complete Road Construction Schemes. Impacts are then usually not assessed for separate roads but for the complete package. Usually these studies are done on demand: the policy makers at the Ministry want AVV to evaluate policy packages. In some cases other parties are involved too in these studies (such as the Central Planning Bureau or the National Institute of Public Health and the Environment).

7.4.3 Defining Scenarios for Use in Policy Evaluation

In co-operation with the Dutch Central Planning Bureau (CPB), the Spatial Planning Agency (RPD) and the National Institute of Public Health and the Environment (RIVM) scenarios for the future of the Netherlands are formulated on a regular basis. Using economic models forecasts for regional economic development are made. The spatial distribution of the houses and the population are based on knowledge of the Spatial Planning Agency of the Ministry of Public Housing, Physical Planning and the Environment.

These new scenarios are used as background material for problem scans and for policy development.

7.4.4 Examples

High Speed Trains. In the Netherlands there are plans for two High Speed Train services: the HST-South, going from Amsterdam-Schiphol to Rotterdam-Brussel and the HST-East, going from Schiphol-Utrecht to Arnhem-Düsseldorf. The consequences of these train services for the demand were estimated using the NMS for the inland travel and other models for the international demand (see e.g. AVV, 1996).

These studies were done in co-operation with Railned, the Marketing Department of Dutch Railways and, in case of the HST-East, the involved Regional Directorate of Rijkswaterstaat.

The so called European Renaissance scenario of the Central Planning Bureau (Centraal Planbureau, 1992; van Wee et al., 1992) was used as the environment for the evaluations. In this scenario the population of the Netherlands is expected to increase from 14.1 million in 1986 to 15.9 million in 2010.[1] The socio-economic data needed at the detailed level of 345 zones for application of the NMS were obtained during the earlier development of this scenario by the Central Planning Bureau in co-operation with AVV.

Car ownership was forecast to grow to a total of 7.8 million cars in 2010, to compare with 4.6 million cars in 1986 using the FACTS model (Pronk and van Wee, 1992). In this scenario economic growth is assumed to be high due to a successful European integration. The car cost per kilometre due to world developments in fuel prices, policy and efficiency improvements increases with 2 % relative to 1990. Both variants with and without a form of road pricing were evaluated. The train fares were also varied, to estimate effects.

The rest of the rail infrastructure was assumed equal to the infrastructure present in 2010 according to the so called Second Tactical Package for Rail Infrastructure Construction, which is the second part of a major investment program in rail infrastructure called Rail 21. The train services running on this infrastructure were based on assumptions made by Railned.

The main conclusions were:

- The influence of car cost, including road pricing on the amount of train passengers is small.
- The influence of train fares on the amount of train passengers is high.
- The influence of the High Speed Trains on inland travel is limited. The estimated growth is largely due to exogenous factors and the assumed quality of rest of the train services.
- The number of international trips by train is expected to increase. The largest part of the border crossing trips that are made using the HST-East were made by train in any case.

Defining Scenarios for Use in Policy Evaluation. In 1997 the Central Planning Bureau (CPB) in co-operation with the Transport Research Centre (AVV), the National Institute of Public Health and the Environment (RIVM), the Spatial Planning Department of the Ministry of Spatial Planning and Environment developed three new scenarios with the purpose to generate information about the relationships between environment, space, energy use and mobility for use in producing new policy documents (such as the Third Transport Structure Plan)

---

[1] Actually this forecast is almost overtaken by reality. Currently the Dutch population amounts to 15.7 million.

(Centraal Planbureau, 1997; AVV, 1997). AVV was responsible for the so called problem scan for the theme *mobility*.

The scenarios were formulated in terms of so called *driving forces*. The following driving forces were distinguished: international, demography, social-cultural, technology and economy. Three scenarios were defined: Divided Europe (DE), European Co-ordination (EC) and Global Competition (GC).

The main differences are that economic growth is high in GC, medium in EC and low in DE, demography is characterised with high immigration in EC, medium immigration in GC and low immigration in DE, technological development and diffusion is slow in DE, medium in EC and fast in GC.

These scenarios were translated to the level of 345 geographical zones that is necessary for application of the NMS.

The purpose of the study was to make a problem scan. That means that only policy measures that are already decided by parliament and no longer subject of debate were taken into account. Investments in road and rail infrastructure are only added if they are in the Budget of the Ministry of Transport for either the building phase or the study phase. The infrastructure assumed for 2010 actually is too large for the financial reservations in the Budget. The resulting infrastructure for 2020, however, fits within the financial boundaries.

Measures that have the objective to reduce the fixed costs of car driving and increase the variable costs are taken in all three scenarios. An extra increase of the fuel prices is implemented in the EC-scenario. Accompanying measures, such as demand management at companies, parking restrictions etc. are assumed to be implemented in a moderate way.

The ratio between public transport fares and average fuel cost per litre was assumed not to change. For freight account was taken of the so called Betuwelijn (rail infrastructure leading from the Rotterdam harbour to Germany and dedicated to freight transport), improvement of the waterways and efficiency improvements in road transport.

Car ownership calculations were made using the FACTS model (Pronk and van Wee, 1992). Forecasts of mobility consequences were obtained using the NMS. Because of the purpose of the study: scanning the future to find potential problem areas, congestion was not fed back to the demand calculations. Resulting developments of freight transport were obtained using the TEM model (NEA, 1992) and the ATTACK model (NEI, 1993). Results of earlier estimations of the effects of the Betuwelijn were used in this study.

The following conclusions can be drawn from the results:

- The car remains the dominant mode in the Netherlands. The expected growth for the car driver mode is larger than the total mobility growth.
- Despite the assumed larger economic growth the car use in the GC-scenario is smaller than in the EC-scenario. This is the result of demographic factors. Trips made for education or with social recreational purposes grow more in the EC-scenario (in which the population grows faster) and trips that are work-related grow faster in the GC-scenario.

- Travel by train is expected to increase substantially. This is caused by increasing quality of the train services until 2010. After 2010 the development of the travel cost is not favourable for public transport, because the efficiency of car engines increases, which causes car cost per kilometre to increase less than the fuel prices.
- In the Second Transport Structure Plan the target for the car mobility is a growth with 35 % relative to 1986. In all three scenarios car traffic increases more than this target, between 45 % and 50 % growth.
- Growth of congestion is underestimated for 2010 and exaggerated for 2020. The realised infrastructure for 2010 will probably be less than assumed (because of budget limitations) and the assumed infrastructure for 2020 is prudent. Furthermore, the reducing effect of congestion on demand is not included in these calculations. Nevertheless: congestion remains far above all set targets in all scenarios. Road pricing was not assumed to be present in these calculations. If road pricing is present, the development of congestion becomes substantially less severe, but the targets are still not met.

Evaluation of Policy Scenarios: The ICES Project. ICES is an abbreviation of the Interdepartmental Committee for Economic Structure Policy. In 1998 by order of ICES the Dutch Central Planning Bureau (CPB) in co-operation with the Transport Research Centre (AVV), the National Institute of Public Health and the Environment (RIVM) and the Social and Cultural Planning Office (SCP) made an analysis of the effectiveness of a package of investments of about 60 billion Dutch guilders in infrastructure, public transport, the vitality of cities, main ports, environment and knowledge. The Transport Research Centre was responsible for the analysis of the impacts of infrastructure investments and investments in public transport. The NMS was used for this analysis.

The economic scenario used for the forecast was the European Co-ordination scenario (EC).

The social economic data for the NMS-zones that were obtained during the scenario development were used in this study.

Reference scenarios were used for the forecasts for 2010 and 2020, in which the extra ICES investments are not spent. The policy that is assumed for these scenarios was based on measures that were already planned at the moment of the evaluation, or for which financial coverage is arranged. Road pricing is not included in the assumed policy for both years.

Two scenarios were evaluated in which (part of) the ICES investments were effectuated: one in which the investments in road infrastructure were effectuated and one in which the investments in public transport were effectuated.

Relevant measures for economic benefits are travel times, vehicle hours lost and total demand (e.g. in car kilometres). These were obtained using the assignment results of the NMS and using the results of the mode and destination choice models.

In the next table the results are shown of the ICES investments in road infrastructure, which amount to a total of 7.5 billion guilders.

Table 7.1. Results of investments in road infrastructure. Percentage difference with regard to reference scenario

Variable	2010	2020
Car kilometres	2.3	3.3
Car speed	0.9	0.8
Vehicle hours lost due to congestion main roads	- 7.3	- 0.2
Public transport passenger kilometres	- 0.1	- 0.4

The results of the public transport investments (in total 15 billion guilders), consisting of extension of rail infrastructure (the so called Second Tactical Package), High Speed Rail, regional light rail projects, and local public transport improvements are shown in the table below.

Table 7.2. Results of ICES public transport investments. Percentage difference with regard to reference scenario

Variable	2010	2020
Car kilometres	-0.1	-0.1
Car speed	0	0
Vehicle hours lost due to congestion main roads	-0.6	0
Public transport passenger kilometres	4.6	4.5
Public transport speed	2.0	2.4

The main conclusions were (Centraal Planbureau et al., 1998; Annema et al., 1998):

- Investment in the main road trunk network is economically sound. Travel time savings represent of course a monetary value that was calculated. In 2020 these time savings amount to 700 million guilders every year. Another benefit is represented by the return of part of the latent demand to the main roads and to peak hours.
- For the concrete infrastructure projects the costs and the relative impact on lost vehicle hours were compared. This led to a priority scale for these infrastructure projects.
- Investments in public transport do not result in substantial modal shift.

- Travel time savings as a result of investment in public transport occur, but the resulting benefits are substantially less than in case of investment in road infrastructure.
- The most effective public transport investments are investments in urban rail. The effects on car use are largest here and the number of travelled kilometres by public transport in these urban areas grows with 9 %.

7.5 Future Improvements of the Model

In the early days of the NMS the emphasis was more than at this moment on estimation of mobility consequences. The emphasis of policy has changed since then towards consequences for accessibility and the effects of policy measures on congestion.

Actually it is very hard to give accurate forecasts of the development of congestion with a strategic model as the NMS is meant to be. The network is not designed for detailed calculations in which the exact length of a off-ramp might influence congestion on the highway or even the presence of a traffic light in the urban area might cause congestion to occur. The assignment technique used within the NMS, QBLOK, is a static technique. Although some of the typical disadvantages of static assignments are diminished through the special algorithm that is used, this can only be done in an approximate manner. The NMS assigns traffic for average hours in three periods, the AM-peak (7.00 - 9.00), the PM-peak (15.30 - 17.30) and the rest of the day which is called Off-peak. There is a Time of Day module that estimates shifts in time of day because of congestion. This Time of Day module does not take account of connections between travel in the AM-peak and in the PM-peak. These two are coupled because of for instance duration constraints of the work activity.

As mentioned before: the predicted amount of congestion strongly depends on the feed back of congestion to the mode and destination choice module. A different way of feed back changes the results substantially and in case of extreme situations there are sometimes difficulties in obtaining equilibrium.

To improve the quality of the forecasts in general, but especially also the ability of the NMS to evaluate measures aiming at congestion reduction the following research is taking place or will probably take place in the near future:

- A project in which new base matrices for the modes car driver and car passenger are estimated, using both flows and surveys in the estimation process (Gunn et al., 1997). In the estimation process congestion is taken into account.
- A project in which the consequences are investigated of using a genuine dynamic assignment in conjunction with the NMS (Gunn and Hofman, 1998; Ben-Akiva et al., 1998; van Vuren et al., 1998).
- The tour frequency models and the mode and destination choice models of the NMS are re-estimated. In the new specification the hierarchical structure of the choice models might differ from the actual specification. An element which

might influence the way congestion, demand and time of day choice interact is the inclusion of an "average level of service for the car driver mode" in the estimation instead of the off-peak level of service that is used in the current model formulation. Unfortunately the inclusion of the inclusive value (or logsum) derived from the Time of Day module is not possible. This Time of Day module predicts only changes in time of day choice relative to the base year and is based on Stated Preference data. The factors necessary for scaling this model to the mode and destination choice model, which is based on Revealed Preference data cannot be determined with the available data.

- A re-estimation of the time of day models. The idea is to improve the realism of these models by clarifying the effect of compensation of cost (in case of a period specific toll), enlarging the number of periods and introducing a coupling between periods with a hazard-based model for the duration of activities. In a later stage the mode and destination choice models can be re-estimated using inclusive values obtained from the new Time of Day choice model.

Acknowledgements

I would like to thank my colleagues Jan van der Waard and Toon van der Hoorn for their comments on the draft version of this paper.

References

Annema, J.A., van Wee, B., van Hoek, T. and van der Waard, J. (1998), 'Evaluation of Dutch public investments plans', presented to the conference *Advances in Integrating Environmental Assessment with Economic and Social Appraisal*, Manchester.

Arentze, T.A., Hofman, F. and Timmermans, H.J.P. (1998), 'Deriving rules from activity diaries', presented to *the 8th World Conference on Transport Research*, Antwerpen.

AVV (1996), 'Vervoerwaardestudie HST-Oost'.

AVV (1997), 'Personen-en Goederenmobiliteit in 2010 en 2020: Prognoses in het kader van de CPB-LT scenario's en de Nationale Milieuverkenning 4'.

Bakker, D.M., Mijjer, P.H., Daly, A.J., Vrolijk, P.C. and Hofman, F. (1994), 'Prediction and evaluation of the effects of traffic measures on congestion and vehicle queues', in *Proceedings of the PTRC Summer Annual Meeting*.

Ben-Akiva, M., Gunn, H.F., van Vuren, T. and Hofman, F. (1998), 'The AIUTO project, implementation of a time dependent assignment at the Randstad test site', presented to *the 8th World Conference on Transport Research*, Antwerpen.

van den Broecke, A. (1988), 'Forecasting long term growth of car ownership in the Netherlands with the COHORT processing model', in *Proceedings of the PTRC Summer Annual Meeting.*

Centraal Planbureau (1992*), Scanning the Future: A Long Term Scenario Study of the World Economy 1990-2015*, SDU publishers, The Hague.

Centraal Planbureau in co-operation with the Adviesdienst Verkeer en Vervoer (AVV), Energie-onderzoek Centrum Nederland (ECN) and Rijksinstituut voor Volksgezondheid en Milieu (RIVM) (1997), *Economie en fysieke omgeving: Beleidsopgaven en oplossingsrichtingen 1995-2020*, SDU publishers, The Hague.

Centraal Planbureau, Rijksinstituut voor Volksgezondheid en Milieu, Sociaal en Cultureel Planbureau, Adviesdienst Verkeer en Vervoer (1998), *Kiezen of delen: ICES-maatregelen tegen het licht*, SDU publishers, The Hague.

Gunn, H.F., Hofman, F. and van der Waard, J. (1993), 'The Dutch National Model for personal travel and its applications in policy analysis', presented to *ORSA/TIMS Meeting*, Phoenix.

Gunn, H.F., Mijjer, P.H., Lindveldt, K. and Hofman, F. (1997), 'Estimating base matrices: The combined calibration method', in *Proceedings of the PTRC Summer Annual Meeting*.

Gunn, H.F. and Hofman, F. (1998), 'TDM simulations in Randstad NL', presented to *the 8th World Conference on Transport Research*, Antwerpen.

van der Hoorn, A.I.J.M., Hofman, F. and Pommer, J. (1995), 'A now well-established part of a consensus society', presented to *the 7th World Conference on Transport Research*, Sydney.

NEA (1992), *Hoofdlijnen TEM*, report no. 920103/12401, Rijswijk.

NEI (1993), *Attack, een Interactief Computer Simulatiemodel voor het Bedrijfsvoertuigenpark*, Functionele Specificatie, Rotterdam.

Pronk, M. and van Wee, B. (1992), 'Market based and technical solutions: The use of the FACTS model in environmental forecasting', paper prepared for *the Cambridge Econometrics Annual Conference: Transport, Communications and the Economy*, Cambridge.

Rijkswaterstaat (1992), *The National Model System for Traffic and Transport*, Transportation and Traffic Research Division.

van Vuren, T., Carmicheal, S., Hofman, F. and Taale, H. (1998), 'Converting static to dynamic assignment models: Preliminary findings', presented to *European Transport Conference*, Loughburough.

van der Waard, J. (1997), 'Dutch experience of using a National Transport Model: The use of the National Model System', presented to *Seminar on National Transport Models*, Noordwijkerhout, The Netherlands.

van Wee, B., van der Waard, J., Blom, U. and Brus, D. (1992), 'Verkeer en Vervoer in drie nieuwe lange-termijn-scenario's', *Verkeerskunde*, vol. 43, no. 6, pp. 16-22.

8 The New Swedish National Model – SAMPERS: System and Validation

Gunnar Johansson Sveder
Transek AB
Solna Torg 3
SE-171 45 Solna, Sweden

8.1 Introduction

In November 1997 the Swedish Institute for Transport and Communications Analysis (SIKA) issued a call for tender to develop a new Swedish national model system. Transek signed the contract in March 1998. The most comprehensive contract in this field ever in Sweden had the ambition to develop the new Swedish national model System, the SAMPERS system.

The scope of the SAMPERS project is i) to develop a user-friendly computer traffic forecasting system, ii) to develop new travel demand models for all trips in Sweden

The aim of this paper is to present the SAMPERS project and the SAMPERS system. The specification of the models and the results from the estimations are not in focus in this paper (see Chap. 9).

8.2 Background

Swedish authorities have a long tradition of developing travel demand models. The first generation of national travel demand models was developed in the beginning of the 1980s, a second generation during the first half of 1990s. These models have been frequently used in a large number of separate planning projects but also in the preparation of the regular national strategic transport investment plans.

The old models covered car ownership, trip frequency, destination choice, mode choice and route choice. The trip frequency, destination and mode choice step was implemented as nested logit models. For long distance trips the trip purpose was divided into private trips and business trips. The regional models were divided into seven different trip purposes. The models were rather comprehensive but not integrated into one single system due to the fact that the models were developed sequentially over the years by different organisations.

One of the main problems with the old model system was the user unfriendliness. A new national travel survey existed which made it possible to update, improve and extend the performance of the traffic demand models.

8.3 The New Swedish National Traffic Forecasting System – SAMPERS

The client had defined more than 100 requirements that the new system should fulfil. The project was divided into six subprojects:

- data quality control;
- estimation of regional model;
- estimation of long-distance national model;
- estimation of international model;
- system development; and
- validation of the model system.

Each of these was then divided into a large number of activities with separate activity leaders. The project had to deliver a system in operation to be used in the work on the Swedish national investment plan to be started in June 1999. To maintain the time schedule the system had to be finished for tests and validation in early February 1999 (deliverable one), less than one year after the contract was signed. Models for international trips, time of day choice and secondary destination were to be delivered during the autumn of 1999.

As subcontractors the following organisations are engaged: Push & Pop data (computer system design), C-net Sweden (data base design), Hague Consulting Company (estimation of international models and international expert group), Swedish National Road and Transport Research Institute (implementation of existing car ownership model) and Deloitte & Touch (adviser to the Transek project management).

Before the project started, the client carried out an investigation of competing traffic assignment systems. The National Swedish Road Administration, the National Swedish Rail Administration and the Swedish Institute for Transport and Communications Analysis (SIKA) developed a "check-list" to investigate the performance of the different assignment systems. Four systems were finally tested by the client, and this effort also included developing general system benchmarks. The four systems were Trips, Visum, Vips and EMME/2. Based on the results, the client decided to use the EMME/2-system as assignment system in the new national model system.

In the beginning of the project, the efficiencies of Fortran and Visual Basic 5 (VB5) were tested. An existing traffic demand model was implemented and the result showed that the processing time was 30% less in VB5, compared to Fortran. Consequently, the system is mainly developed in VB5.

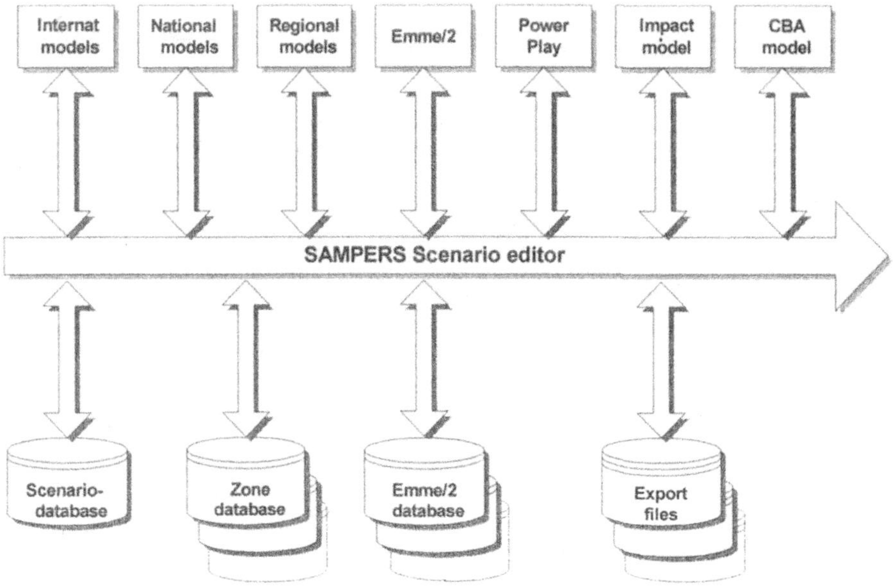

Fig. 8.1. The system architecture

One of the main ambitions of this project was to develop a user-friendly, fully integrated, system. The system should be very flexible for different types of users and different types of projects, and at the same time help the user to avoid implementing errors. A traffic forecasting system operates large quantities of data. The system covers models for 11 different trip purposes, car ownership, impact calculations (for safety, emissions etc) and socio-economic calculations. Altogether the system contains hundreds of parameters and variables which have to be controlled by every user of the system, see Fig. 8.1.

The system is designed for analysing one or several scenarios in each project. The definition of a scenario is a complete set up of socio-economic and demographic zone data, traffic system data and a definition of policy variables, see Fig. 8.2. The owner of the system can define one or several complete set-ups of scenario data and parameters as default. The user can copy the default scenario, make changes of the scenario (traffic system, policy or zone data) and then perform a forecast and an evaluation. To make it easy for the user to control the data and variables of a scenario an information control system has been developed. All data are checked by the check sum and any deviation to default data and variables are identified by the system. Hence the user automatically gets information if any data deviates from a defined base scenario. This information will also accompany the scenario definition and inform any user who reads the results.

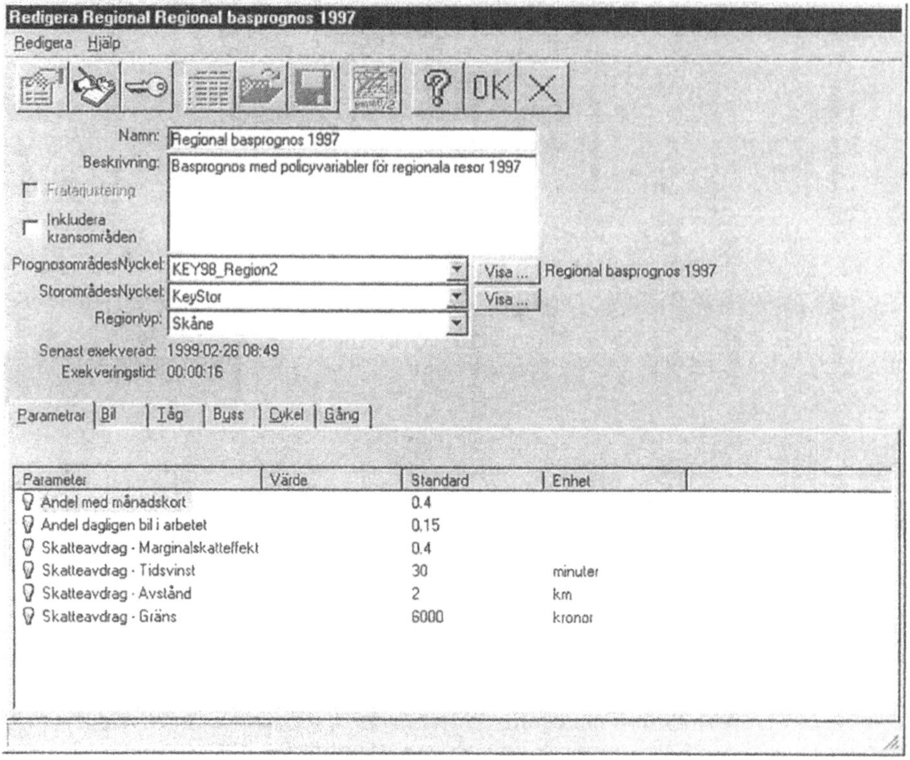

Fig. 8.2. Example of user interface

The following three types of databases are defined; i) zone data, ii) traffic system data, iii) results. The zone data are stored in an Access database. The zone database contains socio-economic and demographic data as the number of people by age class, gender, employment status, and also information on the number of persons employed at different working places and in different economic sectors. The database also handles keys for aggregation of results. Traffic system data, such as travel time components for all public transport modes in the model, travel costs and travel distance are calculated by procedures in the EMME/2 assignment system. The data are stored in an EMME/2 database. The traffic forecasting system reads data directly from the EMME/2 database and the system also writes resulting demand matrices directly into the EMME/2 database. Results as number of trips, traffic mileage, time consumption, accessibility by zones, mode, trip purposes, income group and gender, car ownership and trip distances are stored by the system in a hyper cube. The software Power Play is a flexible tool and is used for analysing and comparing the results in the hyper cube for one or several scenarios and for producing illustrations and graphs, see Fig. 8.3.

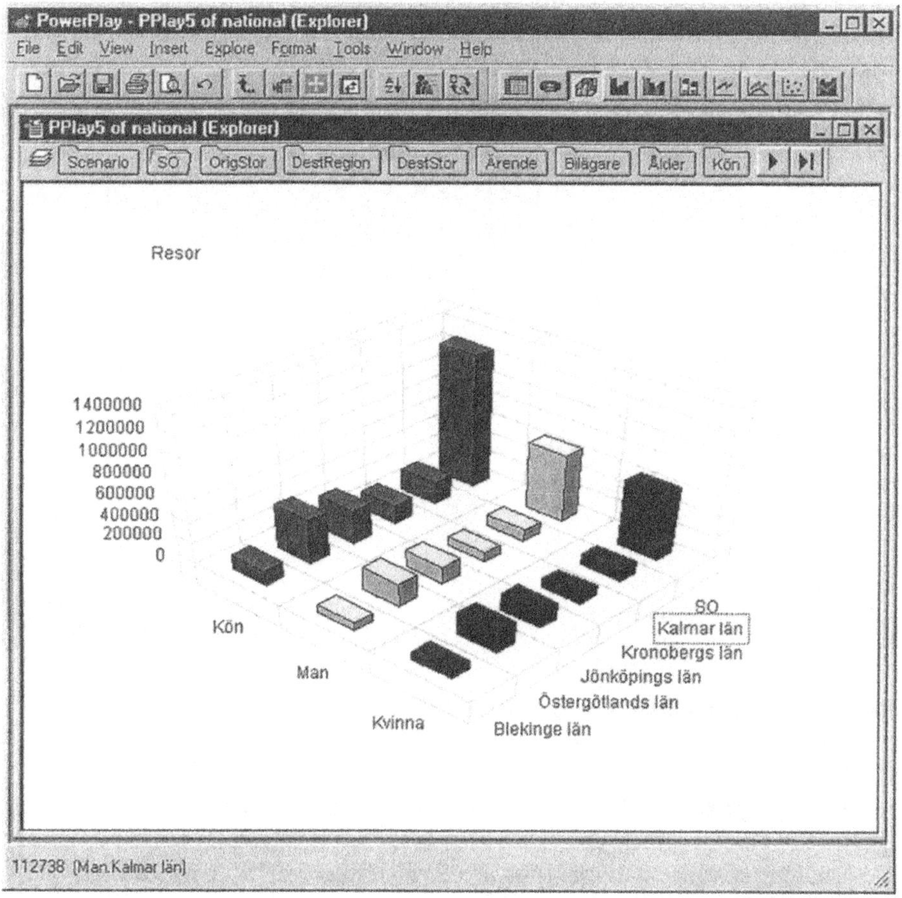

Fig. 8.3. Example of hyper cube evaluation

The models for regional trips are divided into the following trip purposes: work, school, shopping, business and other private. The models are of the multinomial nested logit type and are further described in Chap. 9. The model for the southern region (Skåne) also includes the eastern part of Denmark (Själland), see Fig. 8.4.

The user can decide to run one single model (regional trips by one or several regions, national long distance trips or international trips) or all models simultaneously. The results are assigned to the network by the EMME/2 software. During the model runs the attractions in the zones in the nearest area outside every region are considered, so the destination choice for trips near the region border or across the region border would be modelled correctly.

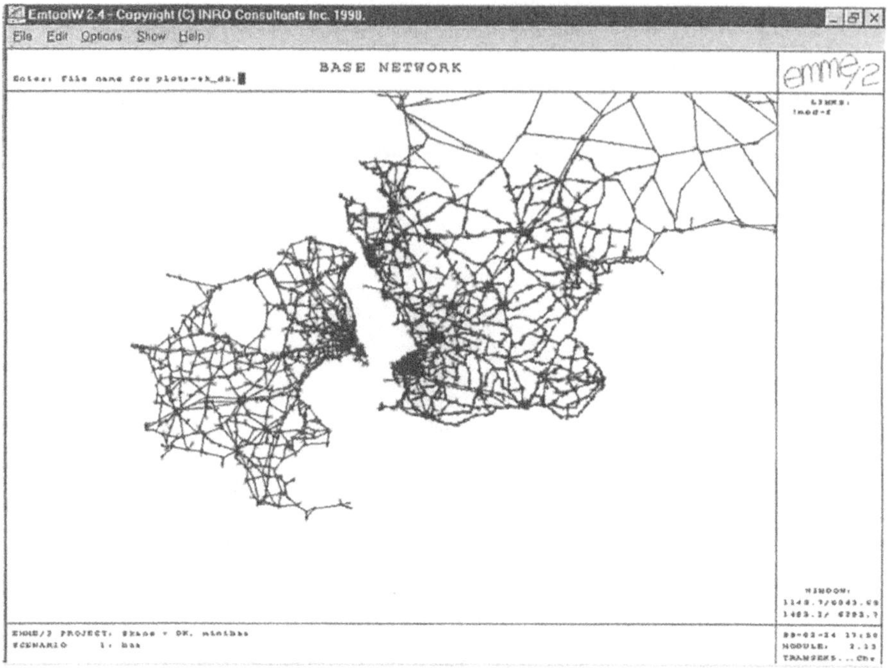

Fig. 8.4. Example of EMME/2 network representation

Some properties of the regional and national traffic systems are summarised in Tables 8.1-8.3.

The long distance national trips are divided into two trip purposes: private and business. They are further divided into journeys lasting one day or longer. The models are of the same type as those for regional trips and are described in Chap. 9. They operate on 670 zones covering all of Sweden.

Table 8.1. Technical specification of the implementation of the regional models

Region	No. of zones	No. of inhabitants	Inhabitants. per zone
South (Skåne)	948	1 100 000	1160
West	2573	1 400 000	544
Southeast	1380	1 300 000	942
East (Mälardalen)	2302	2 800 000	1216
North	1208	1 500 000	1242

Table 8.2. Technical specification of the regional traffic networks

Region	No. nodes	No. links	No. transit segments
South (Skåne)	7749	22 524	19 629
West	7825	30 705	33 061
Southeast	7873	26 510	30 485
East (Mälardalen)	8643	28 709	46 670
North	6804	21 120	23 815

Table 8.3. Technical specification of the national traffic network

	Car	Bus	Train	Air
No. nodes	16402	16515	3526	16515
No. links	39787	38313	2998	411
No. transit lines	-	643	1052	273
No. transit segments	-	43113	13945	289

The international trips are also divided into two trip purposes: private and business. The models will be implemented during the summer of 1999.

The validation of the models includes i) calculation of elasticities, ii) comparison between model results and the Swedish national travel survey (see Fig. 8.5), iii) comparison between modelled traffic demand and traffic counts.

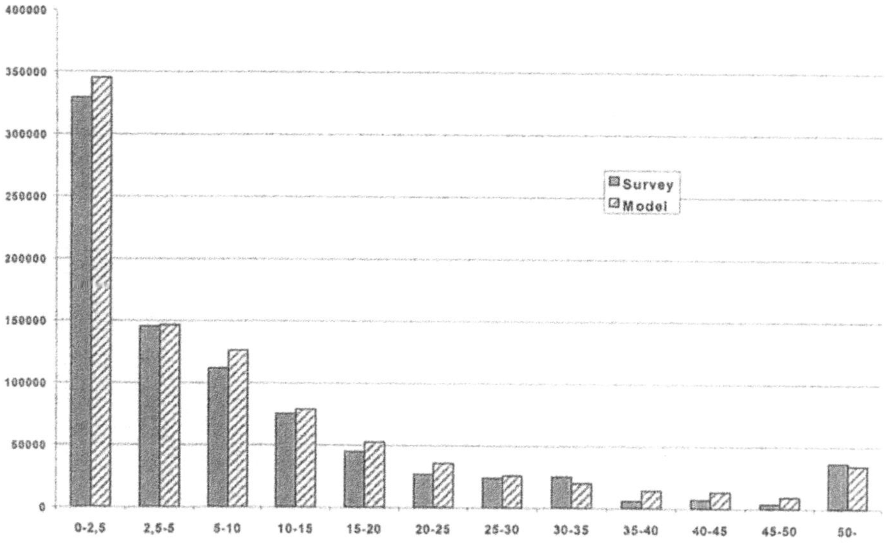

Fig. 8.5. Modelled trips versus survey, regional trips by trip length (km)

A separate model for calculating the external effects of road traffic and other evaluation indicators has been developed and implemented in the system. The model operates on the assigned traffic flows and calculates traffic safety, emissions, energy consumption, and operating and maintenance costs. The calculation considers link type and uses impact models by type of surrounding.

The client had specified the procedure for the socio-economic cost benefit analysis (CBA) and how consumer surplus and external impacts should be valued. The traffic forecasting system saves all relevant data for a standard CBA, and the results can also be compared with a specified base scenario.

The user decides which parts of the traffic forecasting system to apply. The system makes it possible to run a number of scenarios as one single application at one time, including demand modelling (regional, long distance and international trips), impact analysis and socio-economic analysis.

9 SAMPERS – The New Swedish National Travel Demand Forecasting Tool

Muriel Beser
Staffan Algers
Transek AB
Solna Torg 3
SE-171 45 Solna, Sweden

9.1 Background

9.1.1 The Planning Process

Swedish transportation authorities have a long tradition of developing traffic demand models. The first generation of national models was developed in the beginning of the 1980s, a second generation during the first half of 1990s. These models have also been frequently used in a large number of projects but also as a part of the regular national strategic transport investment plan. The national planning process has been a four-year cycle of revising a ten-year investment scheme. The first step in this process is to undertake an analysis to decide on a general policy (like promoting accessibility or focusing on environmental protection). Here the models are used to analyse a few main alternatives, representing major differences in transportation policy and economic development (such as heavily increased petrol taxes to reduce carbondioxide emission).

Based on the decision on the general policy, taken by the parliament, the next step is to perform a more detailed analysis on what projects to include in the ten-year investment plan. The outcome of this process also contains tradeoffs between rail and road investment, which makes it essential to base the analysis on the same forecasting tool and the same assumptions on economic development, land use etc.

The actors in the process are the sector authorities (notably the Swedish National Road Administration and the Swedish National Rail Administration), and a co-ordination authority (the Swedish Institute for Transport and Communications Analysis, SIKA). The forecasting work is carried out by the different actors, and co-ordinated by SIKA. The cost benefit score is a major assessment criterion in establishing the investment plan. Thus it is vital also in this respect that projects in different sectors can be compared on equal grounds.

Finally, the next ten-year investment plan is approved by the parliament. As can be expected, the political process does affect the outcome of the investment scheme.

9.1.2 Cause for New Models and a New Forecasting System

The previously used models covered car ownership, trip frequency, destination choice, mode choice and route choice for long distance, regional and local trips. The trip frequency, destination and mode choice models were nested logit models divided into private and business long distance trips (>100 km) and seven trip purposes for regional and local trips. For route choice and for choice of public transport submodes for regional and local trips, the assignment package EMME/2 was used. The models were rather comprehensive but not integrated into one single system, as a result of them being developed sequentially over the years by different organisations.

One of the main problems with the old models was the lack of integration and their user unfriendliness. A new national travel survey had also been carried out, which made it possible to update, improve and extend the performance of the travel demand models.

SIKA, the Swedish Transport and Communications Research Board and the national planning authorities, all of which forming a group of clients headed by SIKA developed an extensive project specification. After an international tendering process, the project was commissioned to Transek with subcontractors. The main subcontractor for the modelling work was the Hague Consulting Group, the tasks of which included general advice and development of the international models.

A major innovation in the system is a model for long distance trips, extended to include departure time and ticket type choice. This extension is addressed in this chapter, which also reports on the other models developed in the project and on the integration of these models into one system.

9.2 Scope of New System

The general scope of the SAMPERS project was i) to develop a user-friendly computer-based traffic forecasting system, ii) to develop new models covering all trips in Sweden.

9.2.1 All Trips

By the notion all trips is meant trips having at least the origin or the destination in Sweden. Through trips between for example Finland and Denmark are not modelled. As in previous systems, this means that domestic long distance trips, regional and local trips are modelled. The previous models that had been used by the national planning authorities did not contain a model for international trips, so this is a new element in the model system.

9.2.2 Level of Detail

It is obvious that different applications require different level of detail in the forecasting system. An analysis of a road link in an urban environment requires a higher geographical resolution than an analysis of a high-speed train service. Therefore, local and regional trips are handled with a higher resolution than long distance and international trips. For local and regional trips, Sweden is divided into 8,500 zones, which would imply very large matrices and corresponding problems if no further breakdown were made. Therefore, 5 regions are defined, which are run separately.

For domestic long distance trips, 670 zones are defined. The same zones are used in Sweden for international travel. The world outside Sweden (Europe) is divided into 180 zones, coarser as distance increases.

9.2.3 Integrated System

In order to make the model system user friendly, the demand models, the databases and the EMME/2 system had to be integrated into one software under the Windows NT operating system. An important task for the system is also to make it possible to add car trips from the 670-zone level to the regional 8,500 zone-level, and to add up train trips from the regional, long distance and international levels.

Producing a travel forecast is seldom the final step in an investment or policy analysis. To make successive steps in the analysis easier, an effects module and a cost benefit module were required to be included in the system, as well as a module for accessibility analysis.

9.3 Model Overview

The models that have been estimated in the project and which will be more detailed described below are the following:

- regional models,
- long distance models, and
- international models.

All models are of the discrete choice logit type, except for an ordinary least squares trip frequency model for foreigners' travel to Sweden. Experiences from earlier national model studies are also brought into the project (see other contributions in this book).

9.3.1 Data

The main data source for the travel behaviour to be modelled was the national Swedish travel survey, RiksRVU 94-98, which is a continuous travel survey containing 30,000 interviews for the entire interview period. The travel survey contains a one-day diary including all trips, supplemented by trips over 100 km made the last month, and trips over 300 km also made the second last month.

This data set is however not sufficient for international trips, and data collected in other major infrastructure investigations was also used, e.g. the Öresund survey and the Fehmarn Belt survey.

The client using the EMME/2 system supplied the transport supply data. Specifically for long distance trips, data not only for different seasons, days and time of day but also for different years was supplied in order to match the development of the infrastructure over time (such as the introduction of the high speed train X2000).

Statistics Sweden produced land use data. This data was produced at the Small Area statistics level, and then aggregated to the zoning system used in the different segments mentioned above. The number of zones used was as follows:

- local and regional trips 8,500 zones,
- domestic long distance trips 670 zones, and
- international trips 180 zones outside and 670 zones inside Sweden.

9.3.2 Regional Models

General. The task for the regional models is to produce estimates of trips for the following modes: car as driver, car passenger, bus, commuter train, bicycle and walk. The models work on a tour basis. Work tours are defined as home based tours, having a model structure as in Fig. 9.1.

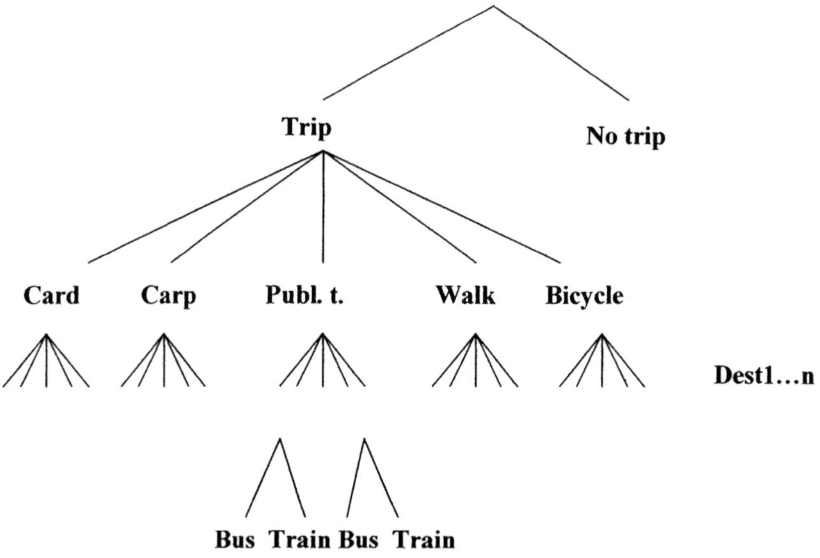

Fig. 9.1. Model structure for home based tours

In order to better reflect similarities of the bus and train alternatives, a structure containing two mode choice levels was adopted, as can be seen from the figure.

Public transport mode choice is now handled at two levels in the model. First, there is a general public transport mode at the mode choice level. Then there is a bus - train mode choice at the lowest level of the model. This mode structure gave a better fit for all trip purposes except for business trips.

There are 5 other trip purposes defined, for which home based tour models have been defined, with the same structure as in Fig. 9.1 (except for business trips, for which too few train trips were reported, forcing the train mode to be omitted form the business trip model). These trip purposes are:

- business,
- school,
- social,
- recreation, and
- other.

The aim of the model structure is also to capture trip chaining when fulfilling this task. This has been done by conditioning secondary destinations and work based tours on the work tour. Thus, if a person has made a work tour, he/she can choose to make for example a shopping trip not only as an ordinary home based trip, but also as a work based tour or as an intermediate stop on the way from work to home. This will of course not capture all types of trip chaining, but a fair share. The structure for non work tours will therefore take the form shown in Fig. 9.2:

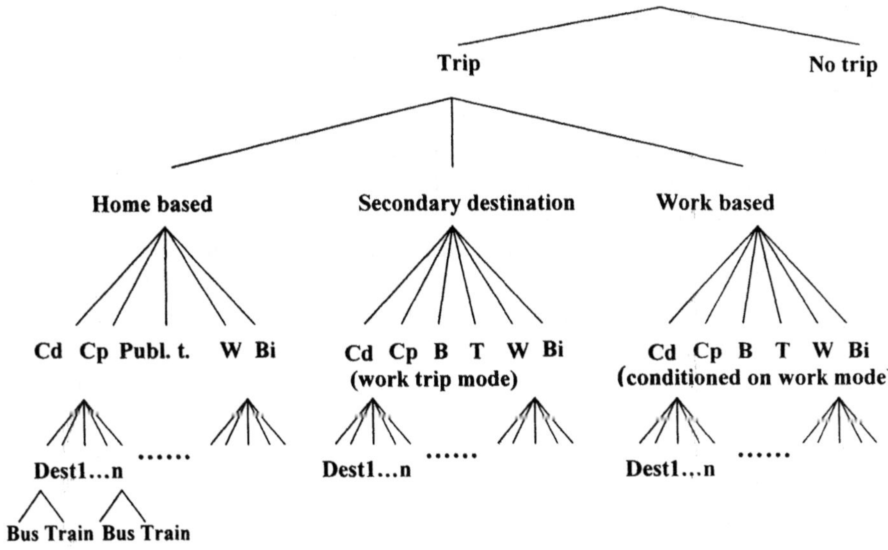

Fig. 9.2. Model structure for non-work tours

Work Tours. A base model was first developed on the total regional data set. Then, the same specification was used on the different data sets for each of the five subregions. A major finding is that the cost and in-vehicle time parameters are very similar in the different regions. Therefore, a model including a number of region specific constants was formulated. In this way, the major part of the regional differences could be accommodated in one single model.

The model was simultaneously estimated, and contains significant logsum parameters between the different choice dimensions.

Concerning travel time components, it has been found that a piecewise linear formulation of (first) waiting time and auxiliary time gives significant improvements. It has also been found that the in vehicle time parameter is very similar for the different modes. An important part of the cost variable is related to Swedish taxation rules, making it possible to deduct work trip travel cost in some cases. The implementation of the model allows for analysing effects of changes in these rules.

The number of destination variables has been very limited. The only size variable used was the log of the total number of employees. Another variables that turned out to be significant was a dummy variable indicating the central area in each county. This variable was defined to be mode specific, and was supposed to account for omissions of parking costs, parking search times etc.

Socio-economic variables have also been included, such as car ownership, license holding, gender and type of employment.

The value of time for work trips amounts to about 55 SEK per hour, which is about 40 percent higher than what was found in the 1994 Swedish Value of Time Study, based on Stated Preference data (Algers et al., 1995).

Non Work Tours. For home based tours, models including mode, destination and frequency choice have been estimated. As for work tours, regional differences were captured by region specific constants rather than estimating one model for each subregion, the main reason being scarcity of data for the train mode.

As for work tours, cost and in vehicle time parameters get significant values. Also, the resulting values of time are substantially higher than those found in the 1994 Swedish Value of Time Study.

Generally, destination choice is a weak part in travel demand models. In previous Swedish models, size variables have normally been defined as the number of employed persons in different economic sectors, because of the general availability of this information. Such variables give of course a very parsimonious characterisation of a destination zone. In this project, an extra effort was made to improve this part of the models. This was done by collecting information on a number of zone attributes, such as number of university students, number of beds in hospitals, number of summer house square meters, dummy variables for supply of winter and summer activities (like skiing facilities and camping facilities) etc. These variables improved the model significantly.

Most of the models contain individuals from 7–74 years. This is of course a fairly heterogeneous group, and the models contain many variables to account for that. It may well be that further segmentation with regard to age would turn out to be efficient.

All models, except for school trip frequency, contain significant logsum parameters integrating the different choice dimensions.

9.3.3 Domestic Long Distance Models

Model Structure. Long distance trips are modelled as tours, longer than 100 km in one direction. The modes for these trips are defined to be car, bus, normal Inter-city (IC) train, X2000 (high-speed) train and air.

A two-phase approach was defined in this case. The first phase concerned the development of a nested disaggregate logit model with frequency, destination and mode choice, which had been used before (Algers, 1993). In the second phase, departure time, class and access/egress mode choice was added to the choices modelled in the first phase.

The model structure of the first phase model is such that frequency choice is at the highest level, mode choice at the middle level and destination choice is at the lowest level. In this model, access/egress was treated in a simple way, by adding a mode specific access/egress distance variable in the main mode choice model. Access/egress was modelled as a regional trip purpose, where the number of access/egress trips was determined by the number of long distance trips. There was no feedback from the regional access/egress model to the long distance model.

Two trip purposes have been defined – private trips and business trips.

First Phase Estimation Results. Both for business trips and private trips, mode and destination choice models have been estimated simultaneously. It turns out that for private trips there is some heterogeneity related to the number of days at the destination. Therefore a further segmentation according to this criterion was made. The first segment model concerns those staying away for more than 5 days, and the second segment model concerns those staying away 5 days or less. The latter model contains a further partial segmentation, where time components are segmented on those who make a day trip, and those who stay away up to 5 days. The specifications are also different in structure – it turns out that the first segment (duration > 5 days) requires the mode choice to be at the lowest level, and that the other segments require destination to be at the lowest level. Thus it seems that we explain destination choice relatively less well for trips with longer duration, which are often vacation trips.

For the destination choice, size variables were first defined as the number of employed persons in different economic sectors. As for the regional models, the extra variables for destination choice improved the model significantly.

Frequency models have been estimated using disaggregate data, but taking the observation period to be less than the maximum trip frequency, thereby allowing the use of observed frequencies in terms of probabilities to make a trip in the defined period (Daly, 1997). As an example, if no one makes more than 10 trips in a month, then a tenth of a month can be taken as the observation period. The probability to make a trip in such a period will range from one for the observed maximum frequency, to zero for those not having made a trip, and values in between for the rest. The advantage of this approach is that the choice becomes a binary choice.

Values of time for domestic long distance private trips for those staying away up to 5 days have been calculated based on the mode and destination choice model.

The values are higher for trips with a shorter duration, probably related to a sharper time constraint. The value of time for trips exceeding 5 days was substantially lower. The in vehicle values of time in the mode and destination model are on the average reasonably close to those found in the 1994 Swedish Value of Time study (which was not segmented in the same way) (Algers et al., 1995).

Also for business trips, values of time have been calculated. As for private trips, trips with shorter duration have higher values of time. As can be expected, time values for business trips are substantially higher than for private trips.

Departure Time and Class Extension. In the standard model, no regard is taken to the timetable. This means, that a situation in which 10 trains are departing evenly distributed over the day is equivalent to a situation in which 5 trains depart in a morning peak, and 5 trains in an evening peak. The difference may of course matter to the traveller, and it may also matter for the track capacity. Therefore, it is important to know when additional departures need to be scheduled to be worthwhile. In order to do this, demand and supply needs to be modelled in a time dimension as well. This was done for air, IC and X2000 trains, but not for car and bus (many bus trips are charter bus trips, which are not scheduled).

Another dimension which is important is the different ticket types which have emerged as a means of price discrimination, typically on air fares, but increasingly also on trains. Therefore, the price for a trip may vary quite a lot, depending mainly on the flexibility needed for the activity in question. In order to account for this, ticket type choice needs to be modelled as well. In this case, a choice between ticket type classes was defined (similar ticket types were aggregated into one ticket type class).

The model structure is thus extended to the two additional choices, and can be illustrated as in Fig. 9.3.

As in the case of the standard model, access/egress mode choice is modelled as a regional model conditioned on the number of long distance trips. The model is the same, and reported in the context of regional models.

Also here, two trip purposes have been defined – private trips and business trips.

Estimation Results – Extended Model. The extended model was estimated after the standard model, and uses the same specification, except for changes implied by the extended structure. The extension implies additional variables related to time choice (difference between the desired departure/arrival time and the one of the time period), and to ticket type choice (booking restrictions, service levels). Such data was not included in the travel survey, and in order to be able to estimate model parameters for these variables, a separate Stated Preference study had to be carried out.

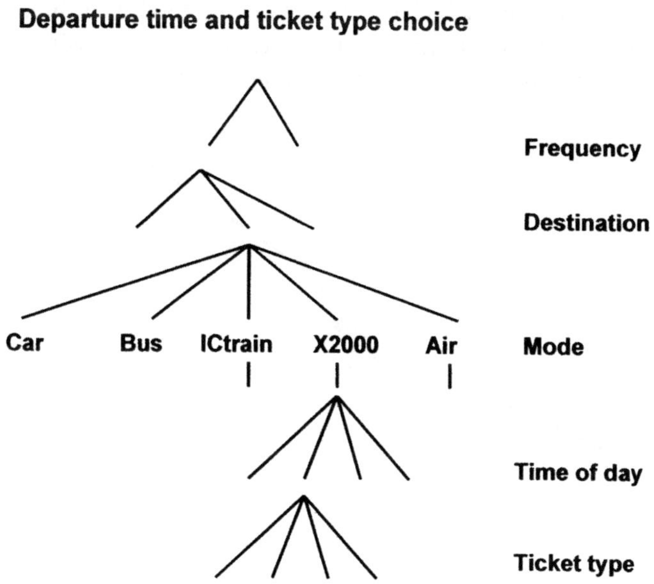

Fig. 9.3. Model structure for long distance extended model

Using parameter estimates from the SP study, fare matrices and trip time components for 30-minute time slices, logsum variables were calculated from the departure/arrival and ticket type levels. In a second step, the models for mode and destination choice were estimated, using these logsum variables in stead of fare and travel time components (which are contained in the logsum variables). The estimated logsum parameters were well estimated, and gave approximately the same fit to the data. The cost parameter, which was generic in the standard model and now is specific to car and bus, increased. The other model parameters were only marginally affected.

The supply for each time slice was obtained by using the Time Table Assignment module of EMME/2 (see Chap. 6).

The values of time concerning the travel time components for air, IC and X2000 train were the same as in the standard model, as they had to be put in at the time of day level logsum variable. For car and bus, they were reduced as compared to the standard model.

9.3.4 International Models

International trips are classified into two main models – one for Swedes travelling to and from other countries, and one for non-Swedes travelling to and from Sweden. There is one important exception from this rule; namely the short trips being made between the very south of Sweden and the Danish island Sjælland (including Copenhagen). These trips are handled as an extension of the regional model for the South region.

Model Structure. The model structure contains three choice dimensions – trip frequency, mode choice and route choice, where routes are classified according to the ferry connection. This structure is depicted in Fig. 9.4 below.

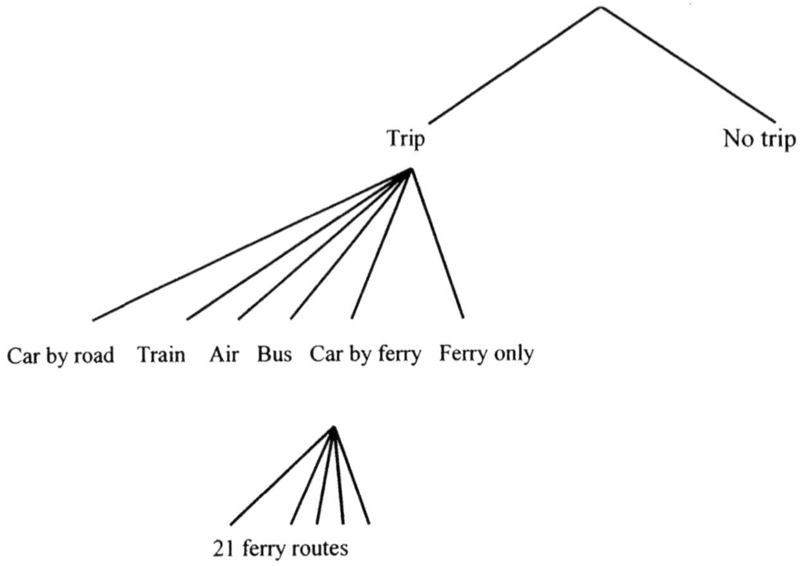

Fig. 9.4. Structure of international trips generated in Sweden

The RiksRVU travel survey did not contain information on the route choice, so other data sources were needed to get the route choice part of the model. A joint estimation of the model is therefore deployed in order to get parameter estimates for the full model structure.

The estimation has resulted in significant estimates for time and cost variables for the mode and route choice parts, and in significant estimates of main variables such as income and accessibility (logsum variables from the mode and route choice parts).

9.3.5 Car Ownership Model

As required by the client, the car ownership model implemented in the system is a previously developed cohort based model for car ownership, based on individual entry and exit probabilities for car ownership. The model was developed by the Swedish National Road and Transport Research Institute. The model gives zonewise car ownership levels. The main variables in this model are income, fuel price, age and company car.

9.4 Validation

The currently implemented models have been subject to some validation. The validation of the models was made in different ways. In the estimation phase, the ability of the model to replicate the choices actually made was tested for different classifications of the data. After implementation, the models were compared with the base information from the travel survey. Finally, the model predictions for the base year were compared to other sources of information, mainly traffic counts. Also, elasticities were calculated before and after model implementation.

These tests were carried out for each of the 5 regions separately. As an example, a comparison for the Mälardalen region (including Stockholm) of the number of tours for different modes is shown below:

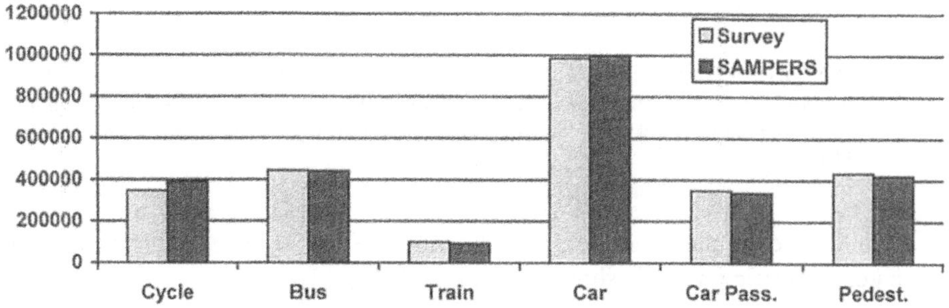

Fig. 9.5. Number of regional tours per weekday in the survey compared to the SAMPERS system by mode (the Mälardalen region)

The trip length distribution for the Mälardalen region is shown in the graph below:

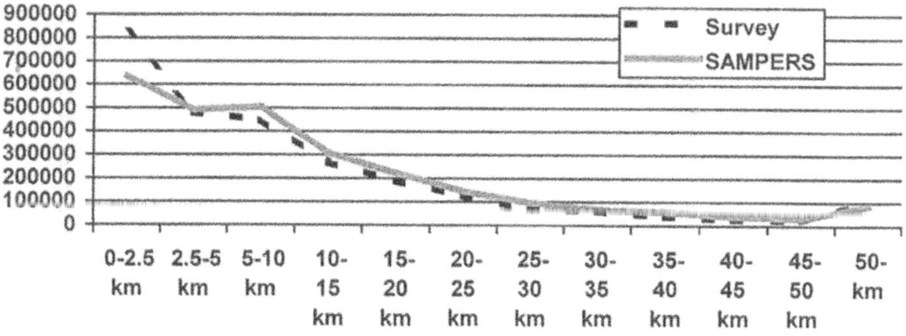

Fig. 9.6. Number of regional tours per weekday in the survey compared to the SAMPERS system by distance (the Mälardalen region)

As another example, the bus in-vehicle time elasticities are shown in the graph below for each trip purpose for the five regions compared to the model before implementation:

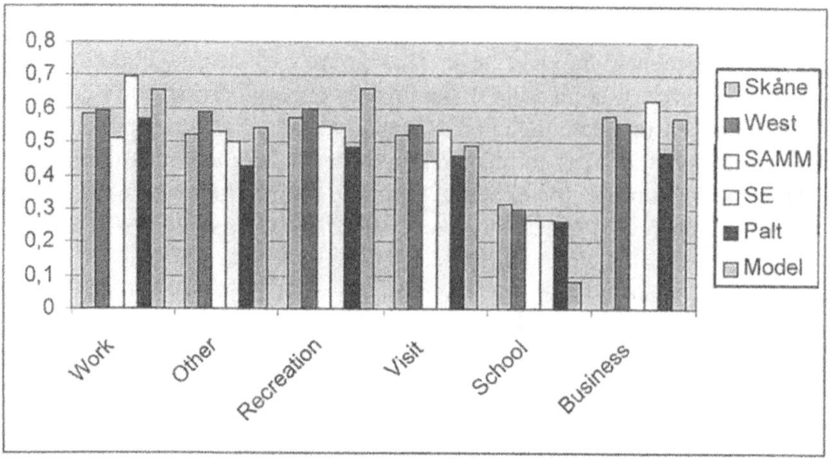

Fig. 9.7. Bus in-vehicle time elasticities from sample enumeration ("Model") compared to the SAMPERS system (the South region "Skåne", the West region "West", the middle region including Stockholm "SAMM", the South East region "SE", and the North region "PALT")

As a final example, a scattergram showing the car volumes after assignment compared to traffic counts is shown (see Fig. 9.8).

9.5 System Design

9.5.1 Windows Menu System

The SAMPERS system is built up as a Windows menu system. The software is developed using Visual Basic as the basic program language. The system contains a number of basic features, that can be put together to forecasting scenarios, which in turn can form a forecasting project. Scenarios are given properties that relate all runs within the scenario to a certain database for zonal data, forecasting year etc. In the same way, projects are given properties that relate all scenarios to the same level of aggregate result output. For forecasting, the following features may be invoked by the user:

- car ownership model,
- the EMME/2 system, directly or by macros,
- regional models,
- domestic long distance models,
- international models,
- disaggregation of trips from national to regional level, and
- iterations (such as car assignment and a regional model run).

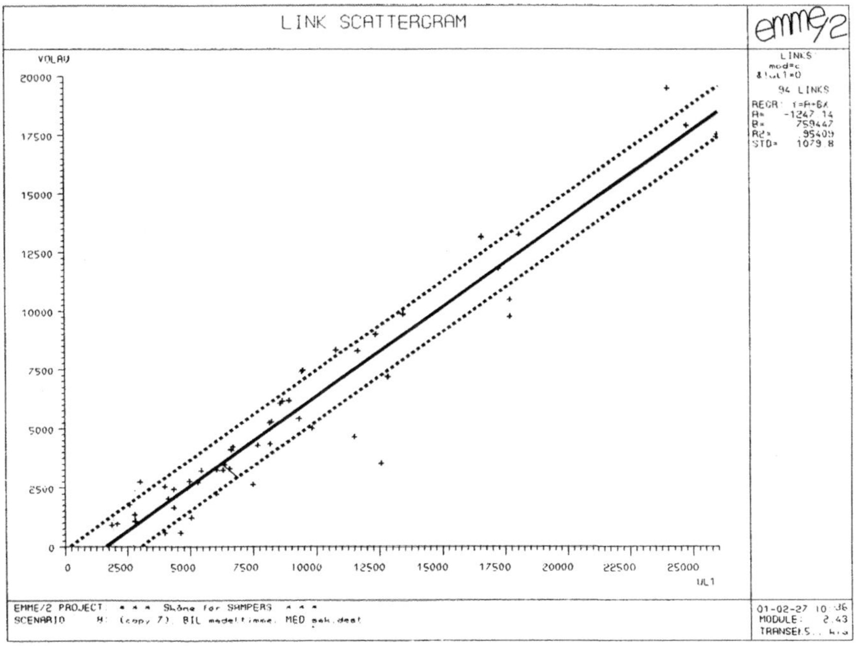

Fig. 9.8. Traffic counts compared to SAMPERS car volumes after assignment, the South region (excluding the city of Malmö)

The main menu is shown in Fig. 9.9, where the user (by clicking on menu buttons) has put together two scenarios, in order to forecast the effects of an extended high-speed train network. The first scenario is the base scenario, and the second scenario is the extended high-speed train network. In the base scenario (which might of course have been run already), the first steps are to calculate supply matrices for the different modes. This is done by invoking EMME/2 macros. Input and output are defined by setting the properties of the macros.

The next step is then to run the domestic long distance model, using the supply data created by the previous macro steps. As for the macros, input and output is defined by setting the properties of the domestic long distance model (as is shown in Fig. 9.10). The last step is then to assign the resulting trips to the network, again by invoking EMME/2 macros for the different modes.

The first step in the extended high-speed scenario is the supply macro for high-speed train, which in this scenario has an extended network. The supply for other modes is not changed, and does not have to be recalculated. The next step is to run the domestic long distance model, now with properties set to match the new high-speed train supply. In a final step, a new assignment is made of the resulting demand matrix for the high-speed train mode.

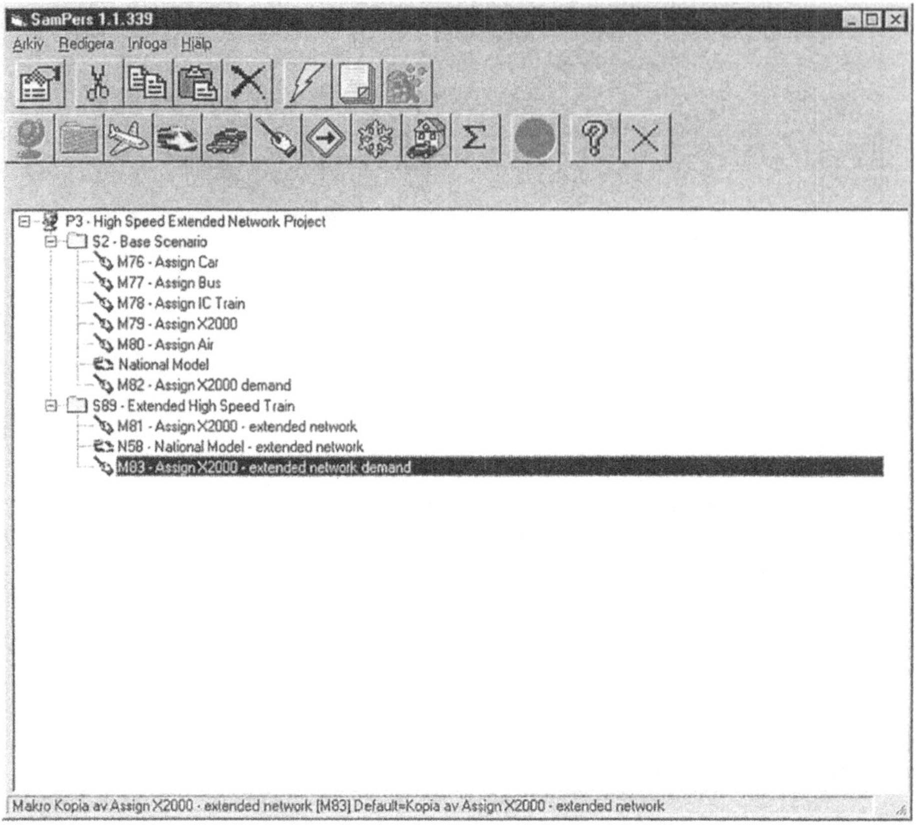

Fig. 9.9. SAMPERS main menu

The definition of the scenarios can be made separately from the actual running of the macros and models, to make the work as efficient as possible. Different steps can also be assigned to different computers, to give complicated runs (such as running many regional models) a shorter turnaround time. Dependencies between different steps can be introduced to ensure that steps are carried out in the right sequence.

9.5.2 EMME/2 Integration

The supply data needed for the models to be run are created in the EMME/2 system. Normally a number of travel time component matrices need to be exported from the EMME/2 databank to be accessed by the forecasting software. Then, result matrices need to be imported back into the EMME/2 databank. This may be very time and storage consuming, and in order to avoid this a more direct process was implemented. Thus, in the SAMPERS system the EMME/2 databank is directly accessed from the SAMPERS modules, for reading as well as for writing.

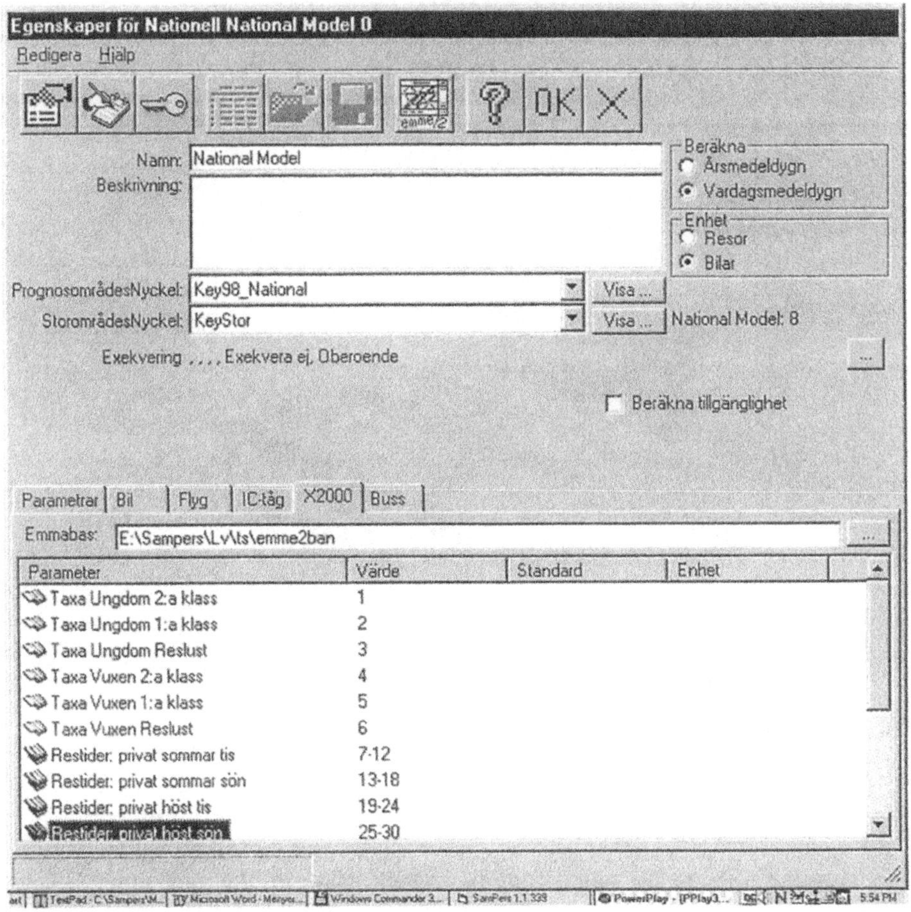

Fig. 9.10. Example of properties sheet for the national model

9.5.3 Results

When the scenarios in the example are run, the user may want to look at the results. For EMME/2 macros, results are stored in the EMME/2 databank, and can be viewed by invoking the EMME/2 system directly from the SAMPERS menu. For model runs, results are always produced at an aggregate level in the form of tables. If the user wants, then also the Cognos PowerPlay system can be invoked from the SAMPERS system, allowing results to be graphically displayed for a rich variety of categories. In Fig. 9.11, the travel demand of a scenario is displayed with regard to mode and trip length.

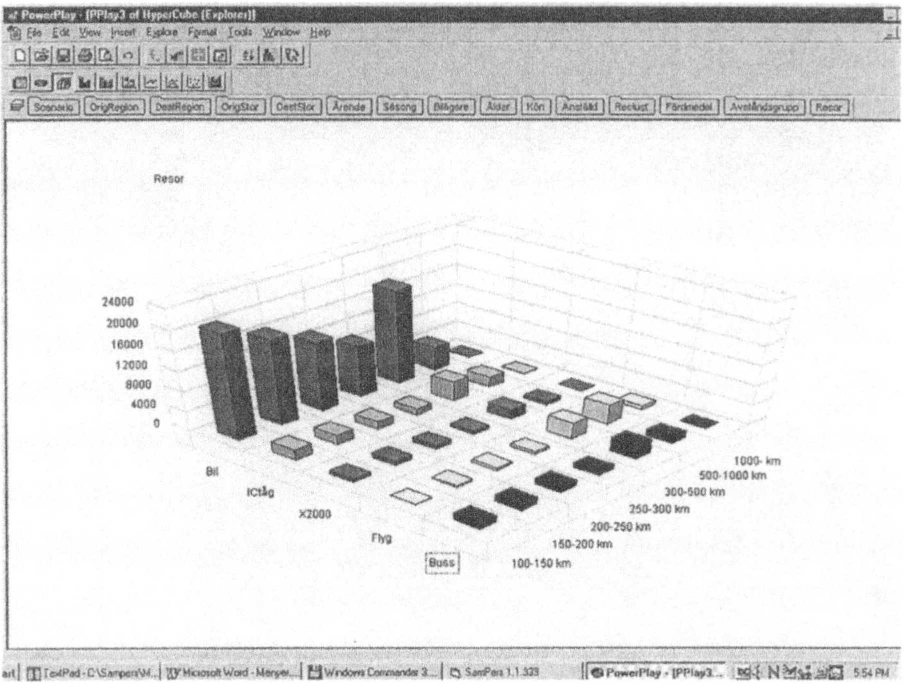

Fig. 9.11. Cognos PowerPlay graph showing results from the national model

9.5.4 Cost Benefit Calculations

The demand forecast is often only a partial result in the planning process. Environmental impact and cost benefit calculations are further steps that are required by decision makers. The SAMPERS system contains two modules to match these requirements. In the effect module, emissions and accident rates are calculated. Then, in the cost-benefit module, these effects are economically evaluated together with the direct costs and benefits in terms of travel time and vehicle cost changes. As the other SAMPERS modules, the effect modules and the cost benefit module are invoked from the menu, and can also be included in the scenario calculations as separate steps like macros and model runs.

9.5.5 Accessibility

Yet another result dimension is formed by different accessibility measures. SAMPERS also contains an accessibility module, where a number of different accessibility measures can be analysed in a GIS-oriented way using ESRI MapObjects (see Fig. 9.12). Some of these measures are related to single scenarios, whereas other measures relate to scenario differences. There are three different types of accessibility measures:

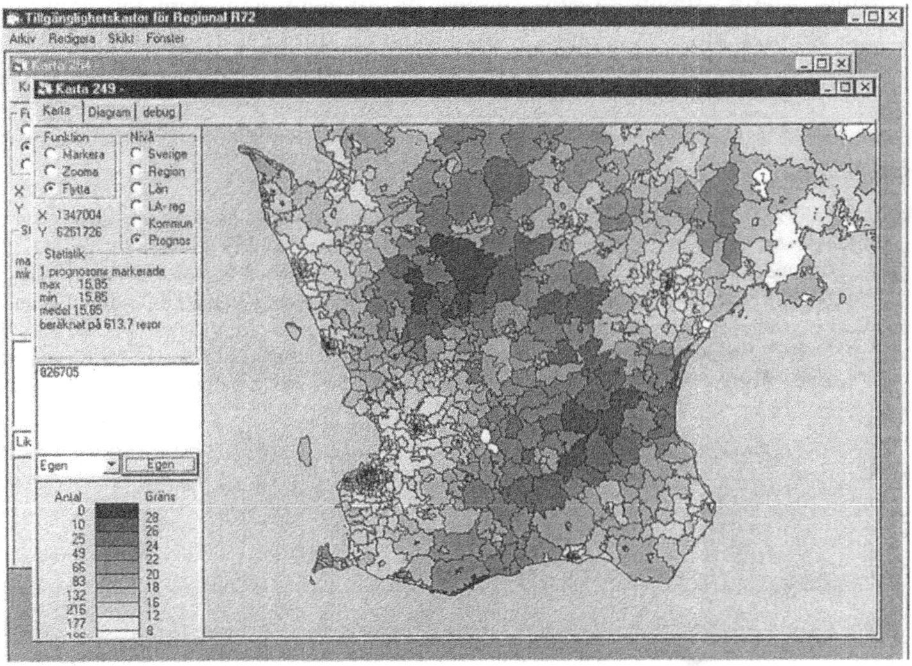

Fig. 9.12. MapObject application in SAMPERS accessibility module

- impedance measures, i.e. travel time or generalised cost to reach certain areas,
- closeness, i.e. how many work places can be reached within a certain time, and
- model based data, i.e. passenger distance travelled, or logsum measures.

9.6 Final Comments

The SAMPERS model system contains many improvements over previous Swedish model systems:

- it covers all personal trip making,
- it is based on recent data covering all trips,
- the model structures are enhanced,
- all models, including assignment, are handled within the same user-friendly software, and
- effects, cost benefit and accessibility analysis can be carried out in the system.

However, as can be expected from a system containing a high level of geographical and socio-economic detail, run times are long. On a standard PC, run times for the regional models (which take the longest time to run), vary from 4 to about 30 hours, depending on the size of the region. If a full set of models is to be run on one computer, several days are needed, especially if many iterations between demand and assignment are needed to capture road network capacity restraints. The possibility to run a project on different computers can however reduce the run times substantially (like running different regional models in parallel).

References

Algers, S. (1993), 'An integrated structure of long distance travel behaviour models in Sweden', *Transportation Research Record* 1413, Washington D.C., pp. 141-149.

Algers, S., Lindqvist-Dillén, J. and Widlert, S. (1995), 'The National Swedish Value of Time Study', in *Proceedings of the 23rd European Transport Forum*, P393, PTRC Education and Research Services, London, pp. 133-149.

Daly, A. (1997), 'Improved methods for trip generation', in *Proceedings of the PTRC European Transport Forum*, P415, PTRC Education and Research Services, London.

10 Italian Models: Application and Planned Development

Francesco Russo
Department of Computer Science,
Mathematics, Electronics and Transportation
University of Reggio Calabria
89100 Reggio Calabria, Italy

10.1 Introduction

In this paper two different national models developed in Italy will be described. The first part (10.2) reports on a large model system for the operational planning of rail services developed by the Italian Railway Company "Ferrovie dello Stato". Section 10.2.1 briefly describes the functional architecture and the main components of the Decision Support System (called SASM) that supports the system model for tactical decisions (Nuzzolo et al., 1995b; Cascetta et al., 1996a). Section 10.2.2 describes the supply model based on a diachronic (spatio-temporal) network representation of scheduled services. Then the demand model is described in Section 10.2.3. It is based on a tree-logit service/run/class choice model that takes into account the desired departure/arrival time. Finally, the assignment model is outlined in Section 10.2.4. It is a stochastic network loading model based on explicit path enumeration.

The second part of the paper (10.3) treats the main developments of a large integrated research project aimed at the construction of a Decision Support System for strategic transport policies (Cascetta, 1996). The project is being financed by the Italian Ministry of Transportation and by the National Research Council (CNR) in the 2nd Finalised Transportation Research Program. Recent developments are concerned with the freight field. Hence, in Section 10.3.1 the freight model system architecture is described. In Section 10.3.2 the introduction of a logistic modal choice model is presented. Finally, in Section 10.3.2 we also report on a behavioural model describing international shipments that modify the traditional approach based on descriptive models.

10.2 DSS for Tactical Planning

The current version of SASM is used by the Italian State Railways to assist decision-making for corridor operative marketing plans for long-distance railway services (SINTRA 1994; Nuzzolo et al., 1995b; Cascetta et al., 1996a; Russo and Beltrami, 1996).

As regards the marketing plan in general terms, the support system is particularly useful in the analysis of the current situation and in the definition phase of marketing strategies by using the following functions:

- supply analysis and comparison of characteristics for railway services and competitive modes (car, bus, plane); and
- effect forecasting, in terms of traffic and returns for the railway system, of variations in service characteristics and service fares.

The decision variables under consideration include the operating characteristics (paths, stops, time-tables and train compositions) and fares of the services. Different tests of the performance of the DSS have been performed (Nuzzolo et al., 1995b) obtaining good results, and the system is currently in use.

10.2.1 The Functional Architecture

Performance of railway and competitive mode/services, passenger "loads" and related returns of each single train for a design scenario are computed through a sequence of phases consisting of (see Fig. 10.1):

- formulation of the design scenario, particularly by the definition of operating characteristics and fares of railway services;
- generation of supply models of the design scenario;
- calculation of alternative runs and relative level of service attributes for demand models;
- calculation of probabilities of each alternative rail/service/run/class configuration for each demand segment and for each O/D pair, by means of a system of choice models;
- application of such probabilities to "current" rail O/D matrices and computation of the passenger number $d_{OD}(s,ser,cl)$ of users of demand segment s on O/D pair, that use rail service *ser* and class *cl*;
- assignment of new O/D flows to networks of the relative railway service;
- determination of the loading rates of every considered train for each belonging link; and
- determination of railway revenues.

The main SASM simulation components are thus as follows:

1) Supply models, that allow the level of service attributes of the various services/modes for O/D pairs to be calculated and the assignment procedures to be carried out:

   a) *Infrastructural road network model and terminals*, consisting of:

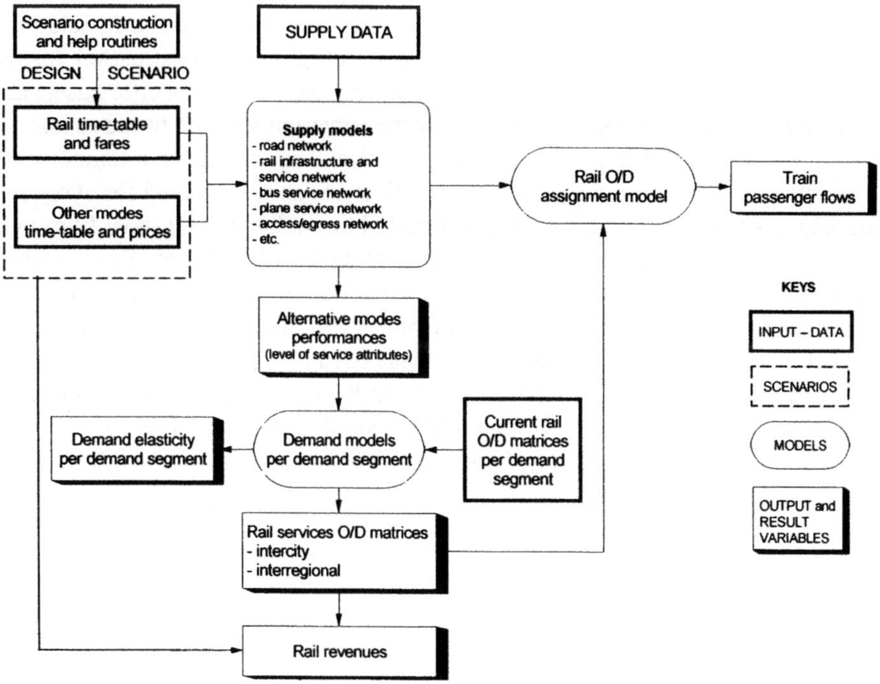

Fig. 10.1. Functional architecture of the Decision Support System SASM

- inter-provincial, inter-regional and national road graphs, with zone centroids and connectors to the network nodes, terminal nodes of railway, plane, bus and ferry, and the relative access arcs; and
- link characteristics, in particular travel time and cost;

b) *Railway infrastructural network model*;

c) *Diachronic network models of railway services* (obtained from train time-tables for intercity, regional/interregional and night services), *air services, interregional bus services and corridor interprovincial bus services* (see Section 10.2.2);

d) *Network model of access/egress from centroids to terminals.*

2) Demand models (see Section 10.2.3);

3) Assignment models (see Section 10.2.4).

10.2.2 The Diachronic Network Model

The network model used (Fig. 10.2) represents each run of each line with an explicit treatment of the variable time; this model is called "diachronic network" (Nuzzolo and Russo, 1994; 1996b; 1997b). The general model of the diachronic graph consists of two subgraphs: one relates to run representation and the other, which consists of various subgraphs, at least one per zone centroid, in relation to the structure of the matrix O/D, segmented by purpose and destination (arrival) target time (DTT) or origin (departure) target time (OTT). In specific cases, as in this model system, it is possible to use only the subgraph referred to the runs.

In the diachronic run graph, for every physical stop k there is a temporal axis on which a node n^r_k, relative to the stopping time of run r, is located for every run r stopping at k (Fig. 10.2). This node is connected through a waiting-boarding link to node n^r_{kb}, which is representative of the departure time of run, through an alighting link from node n^r_{ka}, representative of the arrival time of run r. The link (n^r_{ka}, n^r_{kb}) represents the standing time at stop k of run r; the link (n^r_{jb}, n^r_{ka}) represents the connection between stop j and stop k by means of run r.

Depending on the use, different "costs" can be associated with each link of the diachronic graph, from travel time (corresponding to the difference of time coordinates of the extreme nodes) to a generalised cost given by a weighted sum of time and other disutilities.

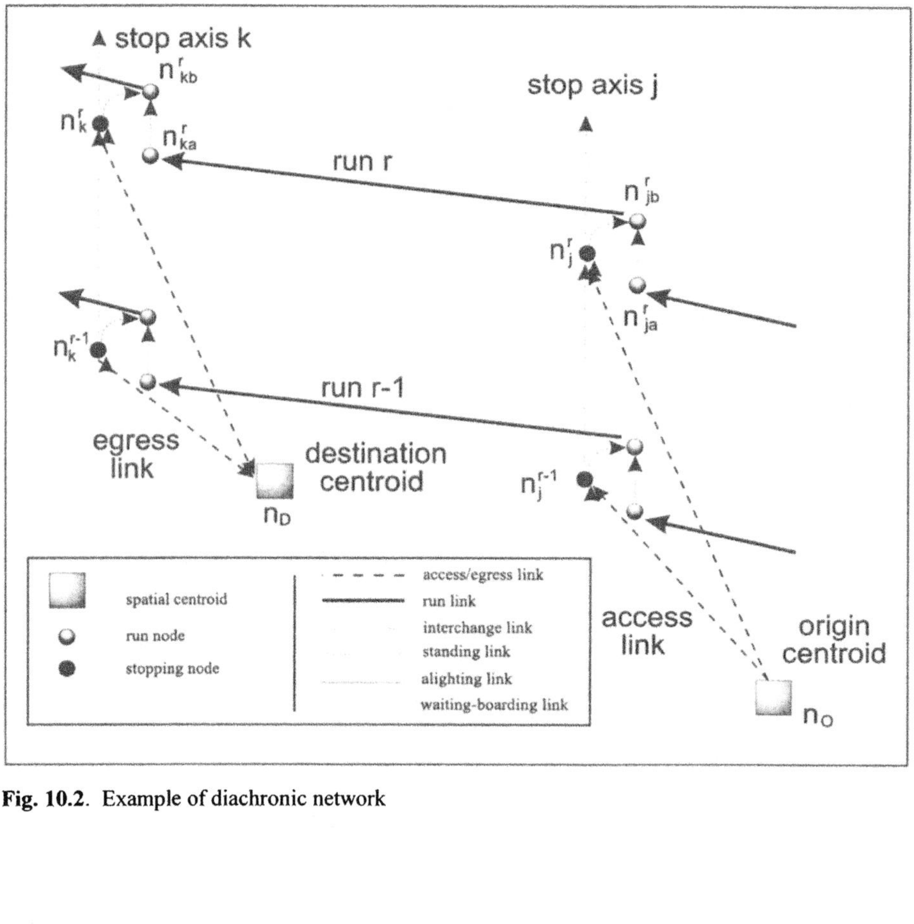

Fig. 10.2. Example of diachronic network

10.2.3 The Demand Model

The demand side of the model system is designed to simulate the impacts of variations in operational service variables (e.g. time-tables, travel times, prices) on user's choices relative to service type, run and class (Cascetta et al., 1996a). The set of these choices can be seen as the equivalent of "route" choice on the multimodal, diachronic service networks and will play this role in the within-day dynamic assignment model described in Section 10.2.4.

The general structure of the model should assume that the rail choice alternatives faced by the user when undertaking a trip (having previously chosen the rail mode), have different dimensions, as shown in Fig. 10.3:

- service;
- run with related access and egress terminals; and
- class.

The interdependence of choice dimensions has been obtained using a tree-logit choice model structure, i.e. assuming a positive correlation among random residuals of perceived utilities of alternatives sharing, service type, specific run, etc. (Ben-Akiva and Lerman, 1985; Daganzo and Kusnic, 1992).

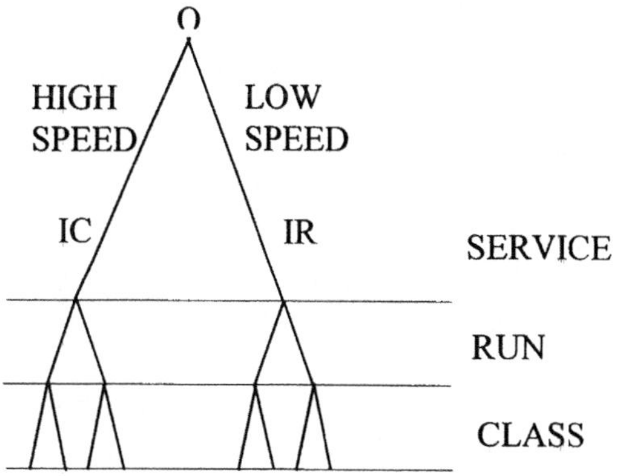

Fig. 10.3. Tree-logit service/run/class choice model (early/late schedule delay approach)

In the used model the probability of choosing a single alternative in the most complex case is given by the equation:

$$p(ser,r,cl/o,d,s) = p(ser/o,d,s) \cdot p(r/o,d,s,ser) \cdot p(cl/o,d,s,r,ser)$$

where:

ser - rail service (intercity, interregional)
cl - service class (I, II)

r - scheduled run including access and egress terminal (first before and first after the target time)
o,d - origin and destination point
s - market segment.

The choice model can be expressed in the factorial form of the following equation corresponding to the tree structure depicted in Fig. 10.3:

$$p(ser,r,cl/o,d,s)=\frac{\exp(V_{ser})}{\sum_{ser'}\exp(V_{ser'})}\cdot\frac{\exp(V_{r/ser})}{\sum_{r'}\exp(V_{r'/ser})}\cdot\frac{\exp(V_{cl/r,ser})}{\sum_{cl'}\exp(V_{cl'/r,ser})}$$

$$V_{cl/r,ser}=\beta_T TT+\beta_{Ch}C_h+\beta_{Cl}C_l+\beta_{eg}\dot{P}_{eg}+\beta_{lg}P_{lg}+\beta_{er}P_{er}+\beta_{lr}P_{lr}+$$
$$+\beta_a Ac+\beta_e Eg+\beta_{Tr}Tr+\beta_{Ic}Ic+\beta_{Ic1}Ic1+\beta_{Ic2}Ic2+\beta_{Ir1}Rg1$$

$$V_{r/ser}=\delta_r Y_r$$

$$Y_r=\ln\sum_{cl}\exp(V_{cl/ser,r})$$

$$V_{ser}=\delta_{ser}Y_{ser}$$

$$Y_{ser}=\ln\sum_{r}\exp(V_{r/ser})$$

where the symbols have the following interpretation:

V,Y respectively, systematic and inclusive utility functions
TT total travel time including transfers (hours)
C_h C_l total travel costs for medium-high (h) and low (l) income class (Italian lire/10^3)
P_{eg} P_{lg} early and late arrival schedule delays with respect to desired arrival time (destination target time DTT) in the outgoing trip, divided by the distance (hours/100 km)
P_{er} P_{lr} early and late departure schedule delays with respect to desired departure time (origin target time OTT) in the returning trip divided by the distance (hours/100 km)
Ac average access (egress) distance for outward (returning) trip (km)
Eg average egress (access) distance for returning (outward) trip (km)
Tr dummy for transfer
Ic intercity service "modal"
$Ic1$, $Ic2$ respectively, first and second class of intercity service "modal"
$Rg1$ first class of regional service "modal"

Calibration of the path choice models have been carried out, using the data of an on-board train survey with about 10,000 interviews on the corridor. For example (Table 10.1), for the business travel the results of the calibration (Cascetta et al., 1996a) give the following utility function (note that, assuming the total variance equal to one it follows that: $\theta_{ser}=\delta_{ser}$; $\theta_r=\delta_{ser}\cdot\delta_r$):

$$V=\beta_1 TT+\beta_2 C_h+\beta_3 C_l+\beta_4 P_{eg}+\beta_5 P_{lg}+\beta_6 Ic+\beta_7 Rg1+\beta_8 Ic1$$

The night service is not enabled in the corridor, so this service has not been considered.

Table 10.1. Coefficient values for "business"

Coeff	*TT*	C_h	C_l	P_{eg}	P_{lg}	*Ic*	*Rg1*	*Ic1*	θ_{ser}	θ_r
β	-4.11	-0.21	-.051	-2.48	-8.75	1.76	0.23	0.48	0.99	0.37
t-St	-1.95	-4.01	-3.06	-2.06	-2.72	2.65	0.83	2.04	0.01	4.21
	rho squared = 0.30						rho bar squared = 0.27			

10.2.4 The Assignment Model

The adopted assignment model can be classified as a within-day dynamic stochastic network loading model with explicit "path" enumeration. The diachronic network model and the system of demand models described in the previous sections allow us to determine the path choice set and the relative attributes for the path choice models.

The path generation and loading procedure for a given railway service, origin and destination pair, a given demand segment and target departure or arrival time, is as follows:

- identification of "feasible" access and egress terminals according to pre-assigned centroid-terminal incidence matrices;
- identification of the set of alternative diachronic run paths;
- linking of diachronic and access (egress) paths;
- computation of choice probabilities considering available alternatives which, in the most general case, include different services/runs; and
- train loading of each alternative.

10.3 A DSS for Strategic Planning of the Italian Transportation System

The DSS is an integrated set of data-bases and simulation models for the national system, both passenger and freight, covering such different aspects as demand, supply, economic and external impacts (Cascetta, 1996). In the general architecture of the system, a central role is played by demand models simulating passenger (Cascetta et al., 1996c; Cascetta, 1998) and freight flows among traffic zones of Italy as well as between Italy and other countries. The main functions of the system are:

- monitoring the Italian transportation system (demand and supply);
- assessment of short and long term policies and support to the Italian Transportation Master Plan; and

– supplying information to operators (local authorities, transport companies, etc).

In the following the system architecture will be recalled for freight transport and the development of logistic domestic and international behavioural modal choice models will be described, representing the main developments of the system.

10.3.1 Model System Architecture

In the Italian project a modelling architecture (Fig. 10.4) combining spatial input-output models (Moses 1955; Chenery and Clark, 1959; Costa, 1973) with random utility mode and path choice has been adopted for reasons of flexibility, data availability and internal consistency.

The multiregional input-output model produces region-to-region trade flow matrices for each economic sector using final demand vectors, technical coefficient matrices for regional economies and interregional trade coefficients which are elastic with respect to the generalised transport cost (Cascetta et al., 1996b). These matrices expressed in monetary terms are transformed into Origin-Destination matrices of freight quantities through specified indices of transformation/amplification (Russo and Conigliaro, 1997).

In the subsequent step the system contains an aggregate descriptive model that splits Origin-Destination matrices of freight quantities, with regard to dimensions, through exogenous coefficients into "captive" large shipments via sea or block train, and unconstrained small and medium size shipments. The O/D matrices of small and medium size shipments were split among subregional zones (provinces) belonging to each region following a gravity model.

Then a macroclass identification model allows aggregation of the quantities of goods belonging to the different sectors obtained by input/output models. It is supposed that there are behavioural homogeneities in decisions regarding freight shipments belonging to the same macroclass of goods. Three macroclasses are currently used with regard to freight: perishables, consumer goods and capital goods.

In the current model system a descriptive size classification model allows us to obtain, for each freight macroclass, the percentage distribution of the whole shipment into a finite number of weight classes. A firm's modal choice model, simulates the distribution among modes of road transport, rail transport, combined transport and shippers (Nuzzolo and Russo, 1996a).

Finally modal O/D matrices are assigned to networks following normative (rail) and random utility (road) path choice models. Consistency among different models is achieved through aggregation/disaggregation procedures and inclusive variables (dotted lines in Fig. 10.4).

The integrated model given a configuration of the transport system allows the analyst to obtain:

– the forecast model O/D matrices;

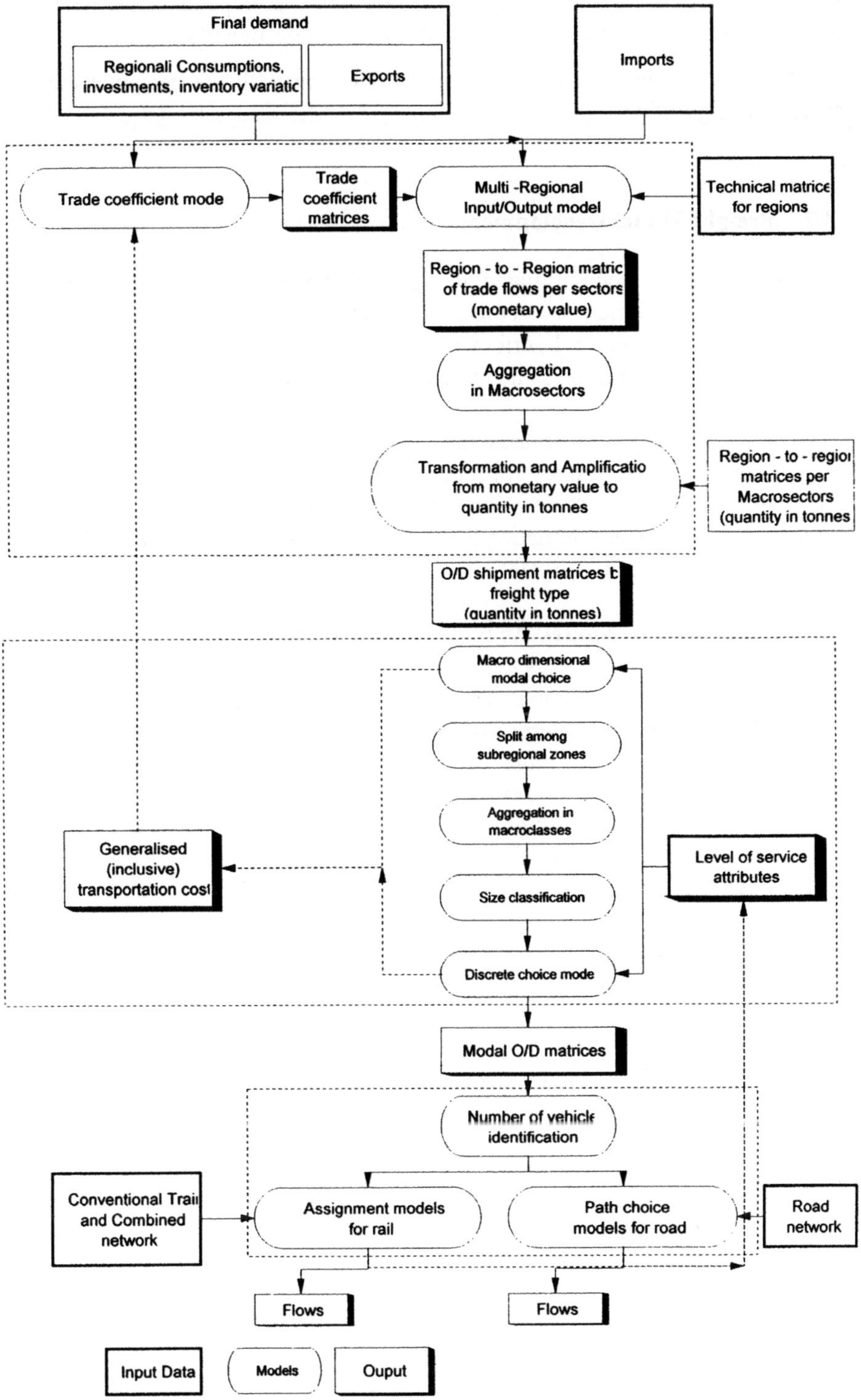

Fig. 10.4. Freight integrated model system architecture of national DSS

- the re-evaluation of the structure of original interdependencies, using the probabilistic model for the evaluation of the trade coefficients; and
- the load of road and rail infrastructures, congruent with the designed transport system configuration.

10.3.2 Freight Model Development

Recent studies have been initiated to develop the discrete choice model in a parallel way: consignment for consumer goods and logistic for capital goods (Bayliss, 1988). This paper describes the firm´s modal choice model in a logistic context, obtaining, within the same model, size and mode chosen (see Nuzzolo and Russo, 1998).

The other development regards the international modal split. In the current model system a descriptive approach (with fixed distribution and modal coefficients) is used to simulate international exchange. Recent work has been carried out to develop an international modal choice model based on a behavioural approach (Nuzzolo and Russo, 1997a).

Logistic Mode Choice Model. As regards the reference unit for the choice model developed in this context, it was assumed that the choice is made within a logistic type approach, with the decision maker being the sender firm.

The set of dimensional alternatives is connected to the modes. As a matter of fact, each mode service has typical load units which allow the minimisation of transport price if used fully in terms of volume or weight; i.e., it is possible to define a specific quantity of freight that can be transported in each load unit. The mode choice set and the utilised load units with the relative average quantities are the following:

Lorry:	Light	(max 20 t)
	Heavy	(30 t)
Traditional Rail:	Two axles	(max 18 t)
	More than two axles	(30 t)
Combined:	Container 20’ (TEU)	(max 15 t)
	Container 40’ (or semi-trailers or swap-bodies)	(30 t)

The disaggregated data refer to a single choice unit such as: $Q_{ds,m}$ global quantities shipped by a single decision maker in a defined time-band (a year was used) of a certain product s between a pair of zones d by a specific mode m. A “random utility” behavioural model was used in order to simulate mode choice for each global shipment, with a systematic utility function of different characteristics, which may be divided into three groups (Table 10.2):

– characteristics of transport alternatives;

– characteristics of sender firms and

– characteristics of the freight to be shipped.

The specification of the logistic utility function involves the need to combine the three groups of characteristics to obtain appropriate logistic attributes. In Table 10.3 the used logistic attributes are reported. It can be noted that unlike other utility specifications, in this case logistic utility has an immediate monetary unit of measurement. In this work we use Italian lire (1 US \$ ≈ 1800 Italian Lire).

Table 10.2. Characteristics of shipments

- Shipment characteristics:
  - ❑ q_m weight of single shipment by mode *m* (tons);
  - ❑ £ value of shipment (Italian Lire/ 10^3/tons);

  Moreover it was assumed that the modal choice for a shipment is conditioned by the following goods classes, called macroclasses:
  - ❑ P_g perishable and consumer goods (0/ 1),
- Firm characteristics:
  - ❑ Q annual weight to ship (tons);
- Transport characteristics (mode m):
  - ❑ t_m. travel time (road, rail, combined) (h);
  - ❑ $c(q_m)$ travel cost (road, rail, combined) (Italian Lire/ 10^3);
  - ❑ PL_m percentage of loss and damage

Table 10.3. Logistic attributes

In transit capital cost	$T_m = \frac{t_m}{365} \cdot Q \cdot £$
Direct shipping cost	$C_m = c(q_m) \cdot \frac{Q}{q_m}$
Loss and damage cost	$LL_m = PL_m \cdot Q \cdot £$
Inventory cost	$I_m = \frac{q_m}{2} \cdot £$

The perceived utility of alternative *j* for user *i* will thus be as follows:

$U_{ij}=V_j(\underline{X}_{ij})+\varepsilon_{ij}$

where V_j represents the mean of the utilities of all users "*i*" who have the same vector $\underline{X}_{ij}$ of attributes. It was hypothesised that the residuals are distributed according to Gumbel function, and multinomial logit models (Ben-Akiva and Lerman, 1985) were estimated, using the maximum likelihood method.

Calibration of logistic model was carried out using data obtained in the research project (Nuzzolo et al., 1995a). In Table 10.4 the first results of the calibration of an explicative model (Nuzzolo and Russo, 1998) for sender firms are reported. Where TR, CO and RO are respectively train, combined and light road model attributes; P is a dummy for perishable and consumer goods in road, and O is a dummy for own rail sidings in rail.

Table 10.4. Coefficient values for explicative logistic model

Coeff.	*TR*	*CO*	*RO*	T_m	C_m	LL_m	I_m	*P*	*O*
β	- 1.96	-1.04	4.15	-0.09	-0.03	-.45E-04	-.21E-04	1.55	0.73
t-St	-2.4	-1.5	4.0	-0.7	-1.5	-1.1	-1.2	1.4	0.5

International Modal Choice. In the current model system a descriptive approach (with fixed distribution and modal coefficients) is used to simulate international exchange. Recent work has been devoted to developing an international modal choice model based on a behavioural approach.

The choice set was specified to include truck, conventional rail service (single wagon) and combined truck + rail service (container, swap-bodies or semi-trailers). A random utility behavioural model was used in order to simulate modal choice for each shipment. The systematic utility is a function of attributes which may be divided into two groups: characteristics of the freight to be shipped and characteristics of modal alternatives (level of service attributes). It is hypothesised that the residuals of perceived utility of alternative *j* are distributed according to the Gumbel function. The set of attributes consisted of:

- travel time (road, traditional train, combined);
- travel cost (road, traditional train, combined);
- border: dummy variables, equal to 1 if foreign countries border the region, 0 otherwise (inserted in road alternative);
- network: rail network kilometres/surface area of country (inserted in rail);
- export: dummy variables, equal to 1 if the shipments are export, 0 otherwise (inserted in combined).

General models and specific geographical area models are specified and calibrated (Nuzzolo and Russo, 1997a) and the main results are reported in Table 10.5:

- General models, termed E*:
  - ❑ Countries with import/export (I/E) of less than 500,000 tonnes were not considered, namely Finland, Ireland and Norway.

- Spain, Portugal and Greece were excluded from the general models owing to problems of interoperability of the railway networks.

– Specific models were calibrated relative to:

- a subset of homogeneous countries with which most I/E traffic traditionally takes place, termed Et: Germany, France, Austria and Switzerland.
- a subset of countries linked by an operational high frequency combined service, termed En: Britain (boosted by the Eurotunnel), The Netherlands, Denmark and Belgium.

Table 10.5. Coefficient values for import/export models

Attribute	Alt.	Unit	E*	Et	En
Train	T	0/1	-0.20*	-3.27	-0.46*
Combined	C	0/1	-0.27*	-1.19	7.50
Train/Combined time	T/C	Hours		-0.006	
Train time	T	Hours	-0.002		-0.04
Combined time	C	Hours	-0.023		-0.32
Road time	R	Hours	-0.001*	-0.001*	-0.12
Cost	G	Lit/10^6	-0.16	-0.16*	-0.22*
Network	T	0/1		48.23	5.33
Export	R	0/1	0.51	0.52	0.26*
Border	R	0/1	0.35	0.31	
Rhosquare			0.21	0.22	0.32
n. Observations			704	408	296

* t-value non-significant at the 5% level;
T = traditional train; C = combined; R = road;
G = generic for T, C, R alternatives.

10.3.3 Conclusions and Further Developments

The main features of the integrated model system consists of the integration of a statistical-economic model, that describes the interdependencies between the regional subsystems of the national economic system, and of a set of models which represent the characteristics of the national freight transportation system and its users' behaviour.

The integrated model is applied within the DSS developed by the Ministry of Transport to assess the effects of the modifications on the configuration of the national transport system, which correspond to different exogenous scenarios, relative to the socio-economic and demographic national and international variables.

The latest developments provide submodels and calibrations of discrete choice models that give the possibility of:

– eliminating the fixed exogenous matrices for modal split of I/E; and

– inserting an explicit analysis of trade-off between size and frequency.

References

Bayliss, B. (1988), *The Measurement of Supply and Demand in Freight Transport*, Avebury, Aldershot.

Ben-Akiva, M. and Lerman, S. (1985), *Discrete Choice Analysis – Theory and Appliction to Travel Demand,* MIT Press, Cambridge, MA.

Cascetta, E. (1996), 'The Italian decision support system for transportation policies and investments: General architecture and development status', in Henscher, D., King, J. and Oum, T.H. (eds), *Proceedings of 7th World Conference on Transport Research*, Pergamon, p. 507.

Cascetta, E. (1998), *Teoria e metodi dell'ingegneria dei sistemi di trasporto*, UTET, Torino.

Cascetta, E., Biggiero, L., Nuzzolo, A. and Russo, F. (1996a), 'A system of within-day dynamic demand and assignment models for scheduled intercity services', paper presented to *the 24th PTRC Summer Annual Meeting*.

Cascetta, E., Conigliaro, G. and Di Gangi, M. (1996b), 'A multiregional input-output model with elastic trade coefficients for the simulation of freight transport demand in Italy', in *Proceedings of 24th PTRC European Transport Forum*, PTRC Education and Research Services, London.

Cascetta, E., Nuzzolo, A., Biggiero, L. and Russo, F. (1996c), 'Passenger and freight demand models for the Italian transportation system', in Henscher, D., King, J. and Oum, T.H. (eds), *Proceedings of 7th World Conference on Transport Research*, Pergamon, p. 473.

Chenery, H.B. and Clark, P.G. (1959), *Interindustry Economics*, John Wiley & Sons, New York, NY.

Costa, P. (1973), 'La regionalizzazione di un modello input-output nazionale: rassegna delle tecniche', *Archivio di Studi Urbani e Regionali*, No. 3-4.

Daganzo, C.F. and Kusnic, M. (1992), 'Another look at the nested logit model', Research report of the Institute of Transportation Studies, University of California at Berkeley.

Moses, L. (1955), 'The stability of interregional trading patterns and input-output analysis', *American Economic Review,* vol. 45, pp. 803-832.

Nuzzolo, A., Frondaroli. A. and Grassi, E. (1995a), 'Un sistema di modelli di domanda e di offerta a scala nazionale', *Preprint of 2° National Conference PFT2-CNR*, PFT, Genova.

Nuzzolo, A. and Russo, F. (1994), 'Departure time and path choice models for intercity transit assignment', paper presented to *the International Association of Travel Behaviour Research Conference*, Chile.

Nuzzolo, A. and Russo, F. (1996a), 'A disaggregate freight modal choice model', in Henscher, D., King, J. and Oum, T.H. (eds), *Proceedings of 7th World Conference on Transport Research*, Pergamon, p. 489.

Nuzzolo, A. and Russo, F. (1996b), 'Stochastic assignment models for transit low frequency services: Some theoretical and operative aspects', in Bianco, L. and Toth, P. (eds), *Advanced Methods in Transportation Analysis*, Springer-Verlag, Berlin, pp. 321-339.

Nuzzolo, A. and Russo, F. (1997a), 'Modal split at international level: A system of models for Italian-based freight', in *Proceedings of 25th PTRC European Transport Forum*, PTRC Education and Research Services, London.

Nuzzolo, A. and Russo, F. (1997b), *Modelli per l'analisi e la simulazione dei sistemi di trasporto collettivo*, F. Angeli, Milano.

Nuzzolo, A. and Russo, F. (1998), 'A logistic approach for freight modal choice model', in *Proceedings of 26th PTRC European Transport Forum*, PTRC Education and Research Services, London.

Nuzzolo, A., Russo. F. and Crisalli, U. (1995b), 'SASM – A DDS for the operative marketing of the Italian Railways', paper presented to *the 3rd EURO Working Group on Transportation,* Barcelona.

Russo, F. and Beltrami, G. (1996), 'A decision support system for the operative planning of railway service on a national scale', in Allan, J., Brebbia, C.A., Hill, J., Sciutto, G. and Sone, S. (eds), *Computers in Railways V*, Computational Mechanics Publications, WIT Press, Southampton, p. 219.

Russo, F. and Conigliaro, G. (1997), 'Integrated macro economic and transport models for freight demand', in Papageorgiou, M. and Pouliezos, A. (eds), *Proceedings of 8th IFAC*, Pergamon, p. 311.

SINTRA (1994), 'SASM – Sistema di analisi e simulazione per il marketing operativo', Progetto esecutivo, Ferrovie dello Stato.

11 PETRA – An Activity-based Approach to Travel Demand Analysis

Mogens Fosgerau
Danish Transport Research Institute
Knuth-Winterfeldts Allé
Bygning 116 Vest
DK-2800 Kgs. Lyngby, Denmark

11.1 Introduction

This paper concerns the PETRA model developed by COWI in a project funded by the Danish Ministry of Transport, the Danish Transport Council and the Danish Energy Research Program. The model provides an alternative approach to activity based travel demand analysis that excludes the time dimension. The approach may be thought of as an extension of the trip based frequency model and allows the model to deal with chains of tours and to treat the choice of activities as endogenous.

The approach has shown both benefits and problems. On the benefit side, the model allows a number of different mode/destination choice models, reflecting different sensitivities for tours with different purposes, interdependencies between trip frequencies for different purposes are meaningfully modelled and a large number of differences between population segments are captured. On the problem side we find that the choice of activities and trip chaining is less cost sensitive than expected a priori. The model only includes core travel activities, which leads to the tentative conclusion that these are less sensitive than non-core activities. Another finding is an income sensitivity of car availability that is lower than found in time series analyses. It is argued that this is, at least partly, due to cohort effects and that time series analyses will overestimate the income sensitivity of car availability.

The model has been applied for a variety of analyses, where an important aspect has been incidence analysis: how different individuals are affected by policy measures. Examples of analyses include an assessment of price and service elasticities for public transport in various urban areas, an assessment of the demand for long distance busses, an assessment of the effects of income tax deductions on commuting, and mobility and taxation differences between urban and rural areas.

The paper is organised as follows. Section 11.2 concerns data, in particular the way trip diaries have been transformed. Section 11.3 provides an overview of the main components of the model system and Section 11.4 presents some results from scenario runs with the model. Finally, Section 11.5 gives some conclusions and discussion of the results.

11.2 Data

The model system is based on more than 13,000 one day travel diaries collected in 1995. One person, the IP, is interviewed concerning his or her travel on the previous day. In addition to travel information for the IP, the survey contains socio-demographic information about all household members. Two persons have been identified in addition to the IP: the head of the household, HP, defined as the adult with the highest income, and the partner, P, defined as the spouse of the HP. The IP need not be identical to the HP nor the P.

11.2.1 Definitions

In order to discuss the way data has been transformed a few definitions are needed. A *destination* is either the home of the person or a location where a person engages in some activity. The activity is called a *travel purpose* or *purpose* for short. A *trip* is defined as the travel from one destination to the next; intermediate destinations where the travel mode is changed are ignored. A *tour* is the sequence of trips that takes a person from home to one or more destinations until he returns at home. Non-home based tours are thus ignored. A *chain* consists of all tours undertaken in a day, each comprising two or more trips between home and destinations. The model is restricted to three possible travel purposes: Work, Errand and Leisure. Each purpose may occur more than once in a chain.

11.2.2 Simplified Chains

We are concerned with a person's travel over a day, characterised by the chosen chain. The observed chains are sometimes extremely complex, necessitating a number of restrictions for chains allowed in the model:

- a chain contains at most 2 tours (92% of observed chains contain at most two tours);
- a tour has one or two destinations away from home; and
- a chain contains at most three purposes.

Then four different travel patterns are possible, see Fig. 11.1.

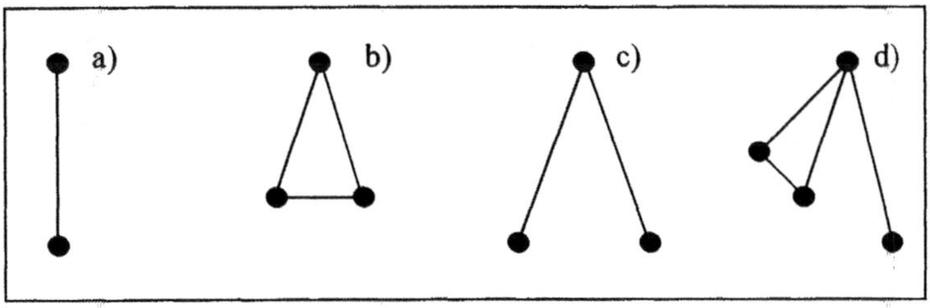

Fig. 11.1. Travel patterns

Next, we make some restrictions on the occurrences of purposes:

- work occurs at most once in a chain (satisfied by 93% of observed chains);
- errand and leisure each occur at most twice in a chain (satisfied by 95% and 97% of observed chains);
- a purpose can occur only once on a tour;
- two tours only occur when work occurs on one of the tours; and
- only one purpose is allowed on a tour not including work

The last restrictions are put into perspective by the observation that 91% of all tours include at most two destinations away from home. The restrictions imply that 12 chains are possible in addition to the empty chain corresponding to staying at home. The possible chains are listed below:

- work, errand, leisure;
- work + errand, work + leisure, errand + leisure;
- work-errand, work-leisure; and
- work + errand-errand, work + errand-leisure, work + leisure-errand, work + leisure-leisure.

11.2.3 Simplification of Observed Chains

A number of observed chains must be transformed in order to conform to the definition above. The transformations always retain a work purpose and destinations further away from home are preferred. We are not concerned with the timing and sequence of tours and can thus freely rearrange tours within chains. Likewise, destinations are rearranged freely within tours. Where necessary, extra trips are added such that the chains always start and end at home. A tour including work is always retained as the *first* tour. Next, the chains are abbreviated such that they contain only one tour in addition to a potential work tour. Of the remaining tours, the one to the remotest destination is selected and the remaining tours are deleted.

11.2.4 Comparison of Simplified Chains with Observed Chains

A number of tours and purposes have been deleted from the data in order to achieve a manageable description of the chains. A comparison of the number of purposes in the original diary with the number of purposes in the simplified diary, reveals an average loss of 34%. The loss is spread quite evenly between the different chain types, with a smaller loss in work related chains and a loss of about 40% in the other chains. The transformed tours are also shorter. Comparing the distances calculated from network data for the transformed chains with the distances reported by respondents we find that the calculated distances sum up to 71% of the observed distances.

11.3 The Model

11.3.1 Model System Structure

The main PETRA components and their interactions are illustrated in Fig. 11.2.

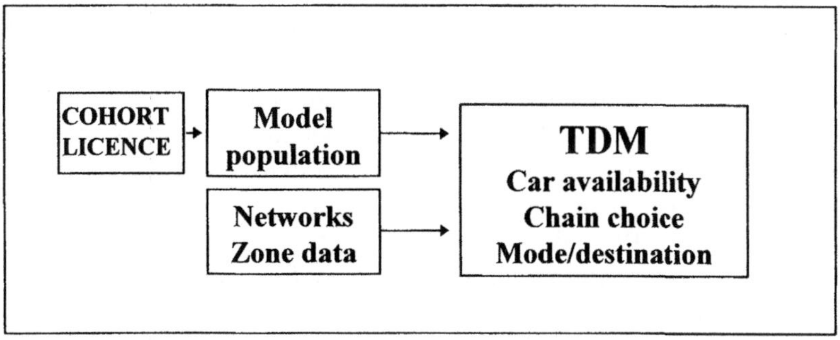

Fig. 11.2. PETRA overview

Cohort or diffusion effects of car ownership result in increased car ownership as the generations change, independently of income growth. These effects are modelled through two satellite models. COHORT models the licence holding rates for groups defined by age and sex and projects these rates from the base year to the forecast year. LICENCE is used to distribute licences on individuals such that the licence holding rates for the model population in the forecast year match those predicted by COHORT. The licence holding data generated by COHORT (COWI, 1997c) and LICENCE (COWI, 1997b) are taken as exogenous by the Travel Demand Module, TDM (COWI, 1997d) as part of the characteristics of the model population. The TDM is applied for the model population using sample enumeration. Using the model population and network and zonal data, the TDM predicts travel demand and the demand for cars.

At the most general level TDM predicts car availability for each household in the model population. For the IP in each household, the TDM predicts probabilities for choosing different travel chains conditional on car availability. At the most basic level, the TDM predicts mode and destination choice for each tour in the chain conditional on car availability and choice of chain.

A TDM run results in a table containing information on the expected travel demand of the IP. The table has the following contents:

- ID, allowing the information in the table to be combined with all the individual and household information in the PETRA database;
- probabilities for the three possible car availability statuses;

- expected kilometres by mode, conditional on car availability; and
- chain choice probabilities, conditional on car availability.

The table is further linked to tables containing weights that reweigh the sample to be representative of the population. Weights are available to make the sample representative at both the individual level and household level.

11.3.2 Mode and Destination Choice Sub-models

A mode and destination choice sub-model has been estimated for each tour type occurring in the chains. For example, chains containing only one work tour, W, are modelled separately. Chains with a work tour and a errand tour, W-E, occur in two sub-models: the work tour is modelled in a sub-model for work tours that are followed by another tour and the errand tour is modelled in a sub-model for errand tours as a second tour. The latter sub-model includes all errand tours made as a second tour, i.e. from chains W-E, WE-E and WL-E.

The sub-models for tours with one destination away from home are specified with 6 alternative modes and 9 alternative destinations, yielding 54 combined mode and destination alternatives. These are set up as logit models, and nested structures with a nest for each mode are estimated. In the case of triangular tours with two destinations away from home, 4 alternative destinations are available for each of 9 primary destinations. This yields 36 alternative combinations of destinations and 216 combined mode and destination alternatives. These sub-models are also set up as logit models and nested structures with a nest for each mode are estimated where possible.

Explanatory variables are time, cost and other level-of-service data for the different modes. Population and employment in the destination zones are used as attraction measures. In addition a number of socio-economic variables for the household enter, notably sex, car availability and licence holding.

11.3.3 Chain Choice

The mode/destination choice sub-models provide probabilities of mode and destination choice, dependent on choice of chain. The choice of chain is influenced, for the different types of tours, by accessibility to destinations and modes. The accessibility of different chains is measured by the logsums from the mode/destination choice sub-models. In addition to the accessibility measures, the model is explained by alternative specific constants for the day of the week plus a number of other socio-economic variables, i.e. sex, age and income.

11.3.4 Car Availability

The logit model on the top level predicts car availability for the household. The model distinguishes between single and two adult households, the latter are referred to as couples. Only few single adult households have two or more cars and few two adult households have three or more cars. Thus, the car availability model only includes the alternatives for 0 and 1+ cars for single adult households and 0, 1 and 2+ cars for couples.

The main variables explaining the model are accessibility measures, income and licence holding, incorporating time dependent cohort effects. Expected car kilometres conditional on car availability are used as accessibility measures. The expected car kilometres are calculated using the chain choice and mode/destination models assuming that the household has none, one or two cars, respectively. The model further employs an urbanisation variable for the residence of the household.

11.4 Scenario Runs

The model system has been run with four scenarios and the results are presented in this chapter. The scenarios are the following:

95B - the base scenario for the 1995 population;
95B+ - as 95B with the fuel price increased by 100%;
20B - the 2020 population, including absolute growth, structural change and the cohort effect;
20IG - as 20B with income increased 64%.

In the two 1995 scenarios, the sample population is scaled to correspond to the actual 1995 population. To obtain the 2020 population, the sample is re-scaled to reflect the anticipated population growth and structural change. Furthermore, the COHORT model has been run to generate average licence holding rates in 2020 for groups defined by age and sex. Next, the LICENCE model has been applied to assign licences to individuals in the household. In the 20IG scenario, incomes are increased by 64% corresponding to an annual growth of 2%. This assignment of licences to individuals is made separately for the 20B and 20IG scenarios as income enters the LICENCE model to explain individual licence holding.

11.4.1 Mode and Destination Choice

The results from the mode/destination choice models have been aggregated to expected kilometres by mode. The aggregation is done by calculating the expected kilometres by mode for each sub-model, conditional on chain choice and car availability, and then forming a weighted average of the distances using the chain choice and car availability probabilities as weights. Table 11.1 summarises the results for the four scenarios.

The changes implied by the scenarios are plausible with the car modes reacting strongest and other modes not reacting enough to offset the change in the car modes. Table 11.2 shows expected kilometres by mode and car availability for the 95B and 95B+ scenarios. The use of car, of course, increases with car availability and decreases with the fuel price. Households with 2+ cars are most affected by the fuel price and use of the car passenger mode (CP) is less affected by the fuel price. Use of all other modes decreases with car availability and increases with the fuel

Table 11.1. Kilometres by mode

	95B	95B+	20B	20IG
Slow	2.0	2.2	1.9	1.7
Car	15.5	10.9	15.5	17.4
CP	4.4	3.9	4.4	4.0
Bus	1.8	2.0	1.8	1.5
Train	0.4	0.5	0.4	0.3
Total	24.2	19.6	23.9	24.8

price. The use of non-car modes does, however, not increase sufficiently to offset the decrease in car use. Thus, the dominant effect of a fuel price increase is to decrease car use while mode shift is much less strong.

Table 11.2 indicates that the mode shift induced by the increased fuel price is highest in 0 car households and smallest in 2+ car households. In 0 car households, kilometres by car decrease by 2.2 while total kilometres decrease by 1.7. In 2+ car households car and total kilometres decrease by 6.0 and 5.8 respectively. Thus, non-car kilometres increase by 0.5 in 0 car households and by 0.2 kilometres in the 2+ car households.

Table 11.2. Expected km by mode and car availability, 95B and 95B+

	95B			95B+		
	0 cars	1 car	2+ cars	0 cars	1 car	2+ cars
Slow	3.8	1.7	1.2	4.0	1.9	1.3
Car	3.8	16.6	21.2	2.4	11.8	15.6
CP	4.8	4.8	3.7	4.0	4.3	3.3
Bus	3.4	1.5	1.1	3.6	1.7	1.2
Train	0.9	0.4	0.3	1.0	0.4	0.3
Total	16.7	25.0	27.5	15.0	20.1	21.7

The fuel price increase thus has a quite strong impact on the expected distances travelled by various modes. According to the model, the primary reaction to a fuel price increase is to choose less distant destinations when travelling by car and the next reaction is to change mode. Increasing the fuel price leads to a decrease in car kilometres of 29.7% in 1995, implying an (arc) elasticity of car kilometres with respect to the fuel price of -0.3. The arc elasticity is lower than the model's point elasticity which is not calculated. The elasticity is at the lower end of the spectrum of -0.3 to -0.5 found in a review by Goodwin (1992). It is, however, consistent with the range given by him when car ownership is assumed to be constant.

11.4.2 Chain Choice

In the TDM, each individual chooses between the 13 possible chains, with probabilities given by the chain choice model. The choice of a travel chain is, in effect, a choice of what travel purposes, if any, to undertake in a day and how to arrange the purposes into tours.

Table 11.3 summarises the predicted number of trips and tours in the four scenarios. An increase in the fuel price has only a marginal effect on the number of trips and tours. It hardly changes the ratio of trips to tours at all, meaning that substitution of triangular tours for two tours does not take place. A change in the population from 95B to 20B, decreases the number of trips and tours and also the ratio of trips to tours. An increase in income increases all three figures such that in 20IG, trip and tour rates are lower than in 95B while the number of trips per tour is almost the same.

Table 11.3. Number of tours and trips per IP

	95B	95B+	20B	20IG
Tours	0.986	0.982	0.960	0.965
Trips	2.118	2.109	2.059	2.074
Trips/tour	2.148	2.148	2.145	2.149

Another way of looking at the changes in chain choice is Table 11.4 that summarises the number of times each travel purpose is met. Thus, in 95B, 41.5% of all IPs have a tour involving the work purpose. The direction of change implied by the scenarios is reasonable, but the changes are rather small. Shifting from the 1995 population to the 2020 population has noticeable effect on the likelihood of a random person of going to work or staying at home. The fuel price increase has only small effect on the likelihoods, while income growth increases the likelihood of travelling to work and errand purposes and decreases the likelihood of staying at home.

Table 11.4. Likelihood of chains including different purposes

	95B	95B+	20B	20IG
Work	41.5%	41.3%	38.2%	38.5%
Errand	33.1%	32.9%	33.3%	33.9%
Leisure	36.6%	36.5%	36.7%	36.7%
Stay home	19.3%	19.6%	20.5%	20.1%

Changing the model population from 95B to 20B has considerable impact on the probabilities of the various chain choices, illustrating that the model does indeed assign quite different choice probabilities to different population segments thus justifying the inclusion of the chain choice model.

11.4.3 Car Availability

The car availability model produces probabilities for different car availability statuses for single and two adult households. In the model, single adult households can have 0 or 1+ cars and couples can have 0, 1 or 2+ cars. Running the TDM with the four scenarios yields the following general results on car availability shown in Table 11.5. From the table it is seen that in the 95B scenario 41.9% of single adult households have a car available and that 18.4% of two adult households have at least two cars available.

Table 11.5. Car availability by household type, percent of households

	95B	95B+	20B	20IG
Single with 1+ cars	41.9%	40.7%	44.7%	60.6%
Couple with 1 car	70.3%	70.1%	71.2%	55.4%
Couple with 2+ cars	18.4%	18.0%	18.9%	39.2%
Total no. of cars (10^3)	1,587	1,563	1,745	2,226
Change relative to 95B		-1.5%	10.0%	40.3%

The predicted response to a fuel price increase is limited but with single adult households reacting most as expected. Total car availability increases by 10% from 95B to 20B, where most of the increase is due to a larger forecast population. The increase in the car availability *rates* follows mainly from increased licence holding rates but also from changed relative weights of different population segments in the population. The increase is largest for single adult households.

The income increase of 64% from 20B to 20IG indicates an increase in the predicted car fleet of 28%, corresponding to an income elasticity of 0.43. In two adult households, the number of cars per household increases from 1.10 to 1.35 while in single adult households the rate increases from 0.45 to 0.61.

11.5 Conclusion

11.5.1 Income

Based on some time series analyses of car ownership in Denmark it is concluded in Brixen (1996) that the long term income elasticity of car ownership probably is around 0.8-0.9. From 1985 to 1995, Danish GDP has grown on average 1.7% per year while car ownership has grown by 1.1% per year, corresponding to a naive income "elasticity" of car ownership of 0.65. The income elasticity found in the present analysis of cross-section data of 0.4 is lower than these elasticities. At the other end of the spectrum we find the Norwegian National Model, (Institute of Transport Economics (TØI), 1992), where car ownership has been analysed in a joint ownership and use model, employing cross-sectional data, revealing an income elasticity of 0.23.

We thus observe that the cross-sectional analyses yield lower income elasticities than do time series analyses. As shown with the cohort model (COWI, 1997c) there are significant cohort effects in licence holdings, and potentially in car ownership. The presence of cohort effects implies that car ownership has increased and will increase independently of income. Jansson (1989) argues that a diffusion process (a cohort effect) must account for a period in Sweden with declining income, increasing fuel prices and still increasing car ownership. He also shows that car ownership has been increasing at given income levels, i.e. car ownership increases independently of income.

Like car ownership, income has been increasing in the past. Thus, there is a danger that what might appear as an income effect in time series analyses, is really a combination of income and cohort effects. If this hypothesis is correct, then the real income elasticity is lower than the one revealed in the time series analyses. The model provides a way of supporting this hypothesis. Comparing the 95B scenario with the 20IG scenario, we find that income growth, population structural change and the cohort effect up to the year 2020, would result in a 40% increase in car ownership, corresponding to an "elasticity" of 0.6, which is close to the time series elasticity. The elasticity is still lower than the elasticity found in time series analyses, although this could perhaps be explained by the higher cohort effect in the past. The cohort effect is almost exhausted by the year 2020.

11.5.2 Chain Choice

The chain choice model of PETRA represents a novel approach and it is therefore of interest to decide whether this extension of the traditional approach is justified by the results. First, we observe that the mode and destination choice models for the different tour types can have quite different parameters. Allowing for such differences enables to TDM to represent the data more closely. To accommodate this, the TDM needs some kind of mechanism to assign individuals to the various mode and destination sub-models. This is a main reason for including the chain choice model. We find that the mode and destination choice models are in fact quite different providing a justification for the complication introduced with the chain choice model. Second, each individual can choose only one chain and thus interdependencies between trip frequencies for different purposes are captured in the chain choice. Third, we find that the chain choice model is able to sensibly account for variation in chain choice probabilities along a number of dimensions. The model thus divides the sample into smaller groups having more homogenous choices of tour type. Hence, the assignment of individuals to different tour types in the chain choice model is meaningful.

In respect of these points, we therefore feel that the chain choice model has justification. However, we find that the predicted response to a fuel price increase is smaller than could be expected. Whether this is a flaw of the model or a result that can provide some insight into travel behaviour is an open issue. Recall that only some of the travel recorded for each IP is included with the model. We include at most one trip to the work place and at most two other destinations. The destinations that are included are those implying the highest distance travelled. To travel to work or to do most errands can be regarded as necessities, as most people would regard many leisure activities as necessary. A plausible hypothesis is therefore that the travel that is included in the TDM is less flexible than the travel that is not included.

11.5.3 Sensitivity of Car Availability to Fuel Price

With the 95B and 95B+ scenarios we have investigated the model's predicted response to a fuel price increase. The fuel price elasticity of the mode and destination choice models is discussed and found to be reasonable. With respect to chain choice and car availability, however, we find that the model is less responsive than expected. Goodwin (1992) concludes that a likely range for the fuel price elasticity of car ownership of -0.15 to -0.30. Oum et al. (1992) in another review find long run cost elasticities of automobile usage of -0.2 to -0.3 for United States and Australia. This implies cost elasticities of car ownership of perhaps half that, since the response to a cost increase is partly reduced usage of each car and partly reduced car ownership. Previous work undertaken by COWI, has assumed an elasticity for Denmark of between -0.12 and -0.30, with the current consensus being that the true elasticity is nearer -0.12.

In comparison to this, we find a fuel cost elasticity of car availability of -0.02 which is rather low. Other models on the same data have also shown very low elasticities. Therefore we feel that the problem with low sensitivity to fuel price changes is not in formulation of the model as such. Rather, we feel it is due to the fact the our data are cross-sectional with no variation in the fuel price. Another problem could arise since we only include some of the observed travel in the model as noted above. There could be other reasons for the size of the responses relating to the use of accessibility measures. The logsums have attractive theoretical properties, that identify them as natural candidates for accessibility measures. However, the logsums are also indirect measures, that reflect not only the cost of mode and destination choice but also the attraction of the available destinations as well as differences in preferences between utilities. Here we have used the expected kilometres by car as the accessibility measure in the car availability model. This yields more significant parameters than using the logsums.

Acknowledgements

The project has been assisted by a steering group, providing a number of helpful comments and criticisms to the project. The steering group members were Staffan Widlert, SIKA, Farideh Ramjerdi, TØI, now KTH, Sweden, Susanne Krawack, the Danish Transport Council and Erik Toft, Danish Ministry of Transport. Also Eric Verroen of TNO has contributed with comments in the initial phases of the project.

References

Ben-Akiva, M. and Lerman, S.R. (1985), *Discrete Choice Analysis - Theory and Application to Travel Demand*, MIT Press, Cambridge, MA.

Brixen, P. (1996), 'Bilestimationer', Unpublished note, Det Økonomiske Råd, Sekretariatet, COWI, Lyngby.

COWI (June 1997a), 'PETRA Synthesis Report - The Forecast Model', COWI, Lyngby.

COWI (June 1997b), 'PETRA Working Paper No 4 - The Licence Holding Model', COWI, Lyngby.

COWI (June 1997c), 'PETRA Working Paper No 5 - The Cohort Model', COWI, Lyngby.

COWI (June 1997d), 'PETRA Working Paper No 7 - The Travel Demand Module', COWI, Lyngby.

Goodwin, P.B. (1992), 'A review of new demand elasticities with special reference to short and long run effects of price changes', *Journal of Transport Economics and Policy*, vol. 26, pp. 155-163.

Institute of Transport Economics, TØI (1992), 'The national model system for private travel, Report on phase II of model development', TØI rapport 150, TØI, Oslo.

Jansson, J.O. (1989), 'Car demand modelling and forecasting', *Journal of Transport Economics and Policy*, vol. 23, pp.125-140.

Oum, T.H., Waters II, W.G. and Yong, J. (1992), 'Concepts of price elasticities of transport demand and recent empirical estimates', *Journal of Transport Economics and Policy*, vol. 26, pp. 139-154.

12 GB Traffic Forecasts – Status and Development

Tom E. Worsley
Russell C.E. Harris
Highways, Economics and Traffic Appraisal Division
Department of the Environment, Transport and the Regions
Great Minster House, 76 Marsham Street
London SW1P 4DR, United Kingdom

12.1 Introduction

This paper gives an overview of the development of national traffic forecasting in Britain. It sets out some proposals for taking this work forward with the objective of setting up a form of national transport policy model. The paper starts by describing the practice of the 1980s, sets out some of the more important changes in national forecasting since then, and considers some possible future developments.

12.2 Background

In the past the Department's National Traffic Forecasts were used mainly as an input into the COBA programme used in the traffic and economic appraisal of road construction schemes in England. The forecasts were published as a range, with the upper and lower bounds the result of high and low assumptions in respect of economic growth and fuel prices, which were the main determinants of traffic growth in the forecasts.

The forecasts, together with other inputs into the traffic and economic appraisals, played an important part in determining the road user economic benefits of individual highway projects. Many of the environmental impacts of these new roads proposals were also estimated using the forecasts, the Environmental Impact Assessment being an important consideration in the decision made by Ministers on whether to go ahead with the project.

The forecasts were revised from time to time to take account of new inputs – for example, a revised fuel price projection – and improvements to the methods used for producing the forecasts. The model was not seen as a means of testing national transport policies. This was partly because politicians had shown little interest in considering national transport policy options and, partly because of this, the structure of the forecasting model was relatively simple. There were, therefore, few sensitivity tests that could be run other than the fuel price/income growth assumptions that were already taken account of in the range of forecast traffic growth.

12.3 The 1980s

Three forecasts were published, in 1980, 1984 and 1989, with similar methodology in each case. The primary determinant of traffic growth was seen to be GDP as a measure of income growth, with fuel price as a secondary influence. Traffic growth was forecast separately for five vehicle categories – cars, light goods vehicles, two categories of heavy goods vehicles, and buses/coaches.

Taking the 1989 forecast as an example, car traffic was forecast as the product of population, car ownership per head and distance travelled per car.

Two models for car ownership were used. The first was based on a survey of household expenditure and income, deriving a cross-sectional relationship between household income and car ownership. However, the growth in average household incomes from one year's survey to the next was insufficient to explain the growth in car ownership, so a time trend was added. This time trend was related in the model to the increase in driving licences. This was roughly linear over the period of the data, but was predicted in the future to approach a saturation level in which 90% of non-retired adults and 50% of retired adults hold a driving licence.

An extrapolatory model had been used in the past, based on a time series regression of car ownership against income, motoring costs and time. This was constrained by a saturation level. In 1989 this relationship resulted in forecasts which showed little sensitivity to the alternative assumptions about income growth used to generate a range of outputs intended to account for uncertainty. An alternative approach was therefore adopted which combined a cross sectional elasticity with one based on time series data.

The published forecast was the average of the results from these two models, justified by general plausibility with reference to international comparisons.

Constant elasticities of car use to GDP and fuel price were estimated and used in predicting growth in miles per car.

The evidence on the growth in light and other commercial vehicle traffic was reviewed. Over time the relationship between growth in income and growth in these categories of traffic had shown significant variation. For forecasting purposes the elasticity of light goods vehicle traffic by GDP was assumed to be one. The elasticity of road freight tonne-kilometres to GDP was also assumed to be one, and this was then divided between larger and smaller heavy goods vehicles based on past trends. Bus and coach traffic (being only 1% of total vehicle kilometres and thus fairly unimportant) was assumed to be constant over time.

This might be summed up as a simple economic forecasting approach to predicting total traffic.

12.4 Supply and Demand

Between 1986 and 1989 traffic on British roads grew by 25%. This was a period of declining real fuel prices, rapid economic growth during the cyclical upturn in the British economy and a high level of consumer confidence. From this high point, which was above the long term trend, and in the light of relatively optimistic economic growth assumptions, the 1989 forecast predicted higher levels of road traffic than any of its predecessors, see Fig. 12.1.

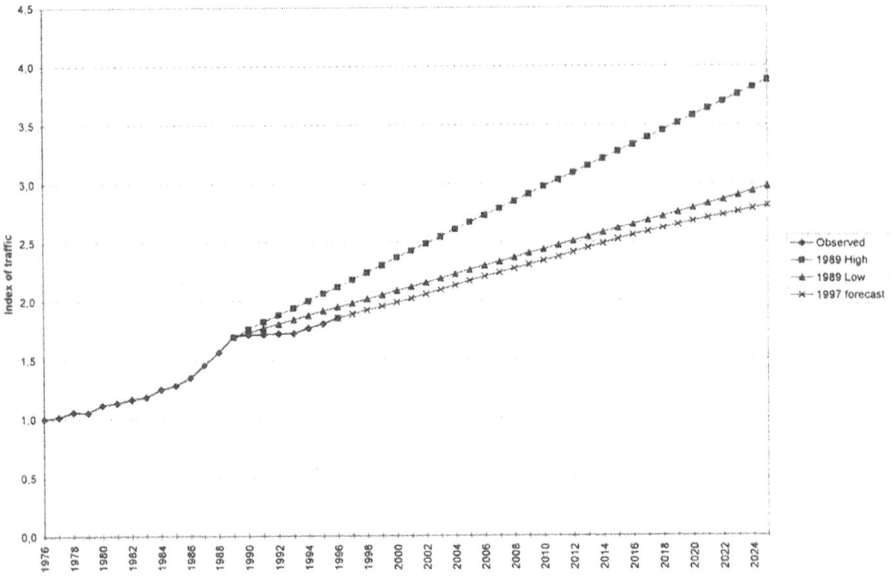

Fig. 12.1. GB traffic forecasts – 1989 and 1997

The forecasts were for traffic growth averaged over Great Britain and were not intended to apply to any specific location.

Indeed, the larger and more complex road schemes were appraised using local models of demand which took into account constraints on road capacity and responses to those constraints. Only the smaller schemes were based directly on the national growth forecasts. However, the absence of any capacity constraints on growth provided an opportunity for groups critical of the Government's road programme to demonstrate that on certain roads and at certain times the forecast growth could not be achieved and hence that the forecasts were based on a conceptual error.

At the same time transport policy moved up the political agenda. Those with an academic or political interest in road traffic became interested in the relationship between road capacity and traffic growth. There was debate as to whether the 1989 forecasts and the roads programme announced when those forecasts were published between them formed a self-fulfilling forecast. Traffic modellers discussed at great length the methods that might be used to reduce forecast traffic growth to the levels that peak hour capacity could cope with, how this should be treated in the economic evaluation of road improvement schemes, and to what extent the missing traffic might be assumed to reappear elsewhere.

Thus it became clear that future forecasting would need in some sense to distinguish supply and demand.

In 1991, a group of officials within the then Department of Transport discussed the desirability of a National Model. It was concluded that this was too difficult to attempt immediately, but might be held as a long-term goal which would guide the Department's research into such paths as would make such a model feasible in the future. The more immediate requirement was to derive some method of constraining forecast national traffic levels to the capacity of the network.

12.5 "Fitting On" and Other Improvements

A six-year program of research and development led up to the 1997 National Road Traffic Forecasts. These incorporated a number of significant advances towards behavioural modelling, including segmentation by household structure. This segmentation allowed for:

- different levels of car ownership by each household type and area type in the base year;
- relationships with income that vary by household structure and area type. Household structures have different saturation levels, which include (where appropriate) the ownership of more than one car; and
- forecasts of changes in the number of households, disaggregated by household structure and by area type in which these changes will take place.

Significant changes were also made to the model for forecasting road freight traffic. The aggregate approach was replaced with one using forecasts of output by main industrial sector, estimates of handling factors, vehicle capacity utilisation and average length of haul.

Forecasts of unconstrained or background growth are derived from the car ownership, car use and commercial vehicle models described above, see Fig. 12.2. Car ownership and car use are not forecast to grow at the same average rate over the entire country in this unconstrained projection. Instead, the forecasts take into account local forecasts of population growth and changes in the number of households to allow for different rates of growth by location and area type within the overall constraint imposed by the nationally based forecast.

A further consideration in the distribution of the unconstrained or background forecasts is the split by road type. The forecasts are based on the assumption that the relationship between national average growth and the growth observed in the past on the five rural area road types and four urban area road types will continue to hold. These unconstrained forecasts of traffic growth by road type which take account of the spatial distribution of population and car ownership growth form an input into the second stage of the forecasting process.

The second stage of the forecasting process requires the modelling of a statistical representation of the supply of road capacity. This procedure was required to ensure that the background forecast was constrained to reflect the availability of capacity. As explained below, this was done by estimating changes in average speeds of this statistical representation of the network and incorporating into the model a range of responses to lower speeds.

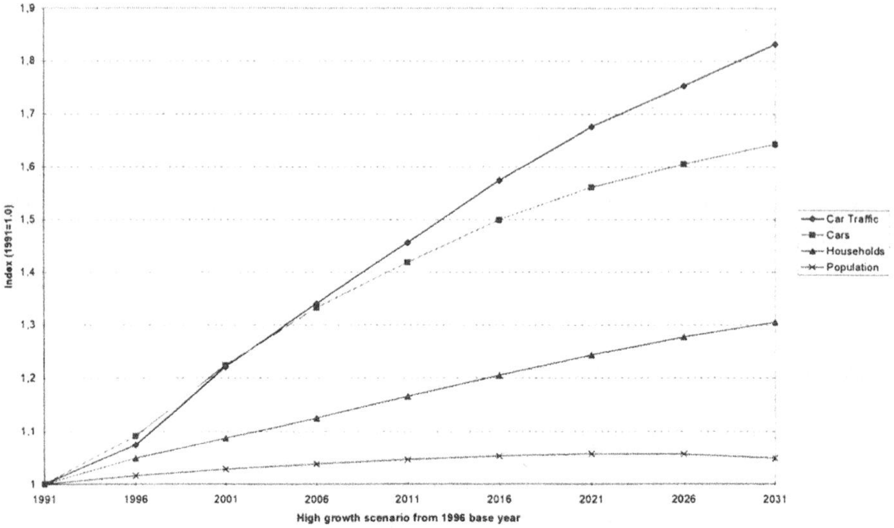

Fig. 12.2. Components of 1997 forecast growth

The first stage of this process was to set up an extensive and detailed database using information collected from recent counts of traffic on the British road network. These data are classified as follows:

- by vehicle type (including different purposes for car trips);
- by road type;
- by region;
- by area type (see Table 12.1);
- by time period;
- by direction (with or against morning peak flow); and
- by congestion level on the road (measured as volume/capacity ratio).

More than 20,000 sites are included in this database, which holds data on traffic levels in the base year, together with information on each of the factors recorded for each of these sites.

Table 12.1. Description of GB in terms of area types

London:	Central
	Inner
	Outer
Other conurbations:	Inner
	Outer
Other urban:	>25 square km
	15-25 square km
	10-15 square km
	5-10 square km
	<5 square km
Rural:	

For each category of road, a speed-flow relationship is provided. For rural areas the conventional speed flow relationships used in the appraisal and design of road schemes are adopted. The urban speed-flow curves take account of the fact that capacity on links in urban areas tends to be constrained by the capacity of junctions. Each of the sites in the database is identified according to the seven factors noted above. The background traffic growth is then imposed on the traffic levels recorded in the database and, where the speed flow curve shows this to result in a reduction in speeds, a hierarchy of demand responses is modelled, which serves, ultimately, to restrain traffic growth.

Research based on stated preference techniques was used to provide estimates of elasticities with respect to changes in speed on the choice of road type/time of travel/distance travelled, see Table 12.2. The research study also provided an estimate of the minimum change in speed or threshold below which no response was assumed to take place.

Table 12.2. Typical elasticities of mileage to journey time change (applied after other switching effects)

Vehicle type / purpose	Elasticity	Threshold
Car commuting	-.14	20% change
Car home-based business	.35	
Car home-based essential other	-.26	40% change
Car home-based discretionary other	-.2	-
Car non-home-based business	-.35	-
Car non-home-based other	-.28	-
Light Goods Vehicles	-.03	-
Heavy Goods Vehicles - rural areas	-.15	-
Heavy Goods Vehicles - urban areas	-.05	-
Buses and Coaches	0	-

The elasticities were applied according to a hierarchical set of rules to determine the effect of increasing congestion on road traffic by road type, region, journey purpose, time of day etc. For example, if speeds fall by more than 20% on motorways in the rural area of the South East region of England during the morning peak (8.00 - 9.00 hours) in the more busy direction of flow, journeys to work will switch:

- to other less congested motorways in that direction/time period/area (subject to a limit of 10% of traffic switching to reflect the inflexibility of land use patterns);
- to the shoulders of the peak on that link or another link of that road type;
- to a different class of road in the relevant area type if it offers a higher speed during the peak or shoulder; or
- to a different mode or destination, resulting in mileage suppression.

When, after the full iterative process has been followed through, traffic flows on a given link and in a given time period remain above the maximum capacity of that link, this flow is simply cut off and assumed to be suppressed.

The representation of the road network is purely statistical. No network model is involved. Despite the large number of sites included in the database (in excess of 20,000), there are limits on the extent to which the results can be disaggregated reliably. There is not sufficient information on, for example, urban area type 8 – the 60 urban areas in England with an area of between 10 and 15 square kilometres – to allow for these to be disaggregated by region. The capacity of the network and existing levels of traffic, split by road type, time of day, vehicle type and journey purpose must be taken as typical of those area types as a whole. Any analysis of policy options for this category of urban area is inevitably relatively general and does not relate to any specific town.

12.6 The Policy Context

In May 1997 a Labour government was elected in Britain. Within a few months of taking office, it issued a consultation paper on transport. This invited individuals, local authorities and other groups with an interest in transport issues to suggest how transport policy might better achieve the Government's sustainable development objective. Sustainable development combines steady economic growth with increased prosperity and jobs for all, providing better protection of the environment and ensuring that all members of society benefit from new policies.

By this time the development of the national road traffic forecasting methods described above was complete. The new forecasts were provided on the basis of the policies in force when the new Government came to power and these were published as a contribution to the consultation process. The forecasts demonstrated the growth in traffic that would take place in the absence of new policies to restrain growth and encourage the use of other modes. The response to the consultation and extensive discussions between Ministers and their advisers were brought together in a Transport Policy White Paper. This document "A New Deal for Transport;

Better for Everyone" sets out the policies that the Government will follow; it is supported by analysis of the likely impacts of these policies where such analysis could be carried out. The National Road Traffic Forecasts are used in the document to set the background and show the general impact on congestion and the environment if policies and behaviour were to remain unchanged.

A more significant development was the use of the forecasting process as a policy appraisal tool. The White Paper announced that the Government would introduce legislation to enable local authorities to introduce congestion charging and retain the revenues to fund transport improvements and other local needs. The NRTF model was used to provide Ministers with information on the likely impacts of congestion charging if applied extensively in English cities. Estimates for typical levels of road traffic reduction were taken from the small number of congestion charging studies that had been carried out in GB. These typical traffic reduction estimates were imposed on the appropriate NRTF area type – for example, large and medium sized cities – to assess the impact on traffic speeds by time of day, road type etc. The model was enhanced to provide information on changes in generalised cost in place of speeds (by including the appropriate journey purpose weighted values of time) and changes in emissions, to increase its value as a policy tool.

The model has also been used to estimate the impact of alternative fuel tax/price assumptions on forecasts of road traffic's contribution to CO_2. This analysis was used in the Government's Consultation Paper on Climate Change, published in October 1998. The model has a number of advantages on the conventional approach based on the direct price elasticity response alone. These include:

- taking account of changes in speeds which follow from the fuel price induced changes in traffic levels;
- providing, if required, a measure of the welfare costs of fuel price increases intended to reduce global emissions by providing an assessment of:
  - the increase in welfare costs on account of the higher fuel tax – the simple welfare loss triangle;
  - the reduction in welfare costs because of lower levels of congestion.

Work is also in hand using this framework for estimating the impacts of measures intended to improve local air quality in urban areas.

This use of the NRTF framework for policy testing has significant limitations. It is not intended to show what might happen in any specific location. This is not necessarily a disadvantage; Ministers do not know, at present, which English local authorities might come forward with congestion plans and so the model serves to show broadly the effects of, for example, half or one quarter of authorities introducing such plans.

A more important limitation is the way in which the change in traffic levels is supplied. Although the model distinguishes between central, inner and outer London and between the inner and outer areas of the five major conurbations, more specific zonal detail is not available. Moreover, the system does not at present allow for redistribution of trips to other area types in response to urban transport policies.

12.7 Plans for Model Development

The demand from the politicians is now for greater policy sensitivity in the forecasts, so as to better estimate the impact of the various measures proposed in the integrated transport White Paper (published in July 1998). But they also in time to inform the setting of national traffic growth targets and air quality targets. The time scale precludes anything but an evolutionary development of the existing forecasting procedures.

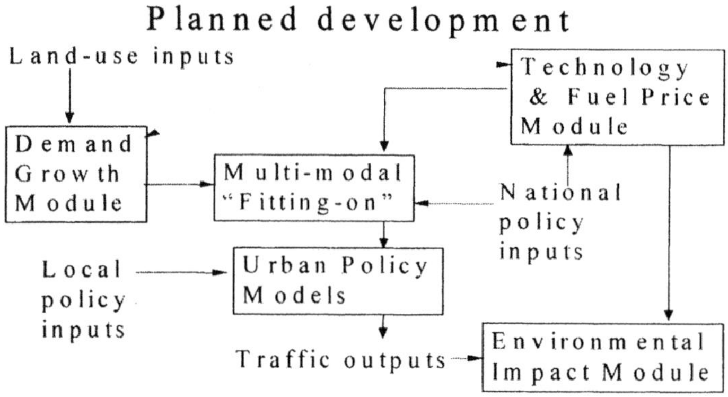

Fig. 12.3. Planned model development for policy sensitivity

Several major areas of work are envisaged, so as to take account of the whole range of policy initiatives under consideration (see Fig. 12.3):

- better modelling of *demand growth*, in particular forecasting trips by all modes, and linking housing and employment growth in different types of urban or rural area to changes in modal split and average trip length;
- reconciling the forecasts of average *trip length* generated by or attracted to each area type with the information on road traffic in the road traffic database;
- looking at *vehicle technology and fuel price* so as to model the spread of fuel-efficient or low-emission vehicles through the car fleet under different pricing assumptions, and link this to the mileage driven by such vehicles;
- reviewing the *speed/flow* assumptions, in particular as applied to urban areas;
- linking in *urban policy models*, to look at city centre charging, parking, and other policies likely to be adopted by local authorities, to replace the very simple approach used at present;

- better *national network* data in the base year, to ensure that the fitting-on process is robust down to finer levels of geographic detail;
- development of the fitting-on process to make *generalised cost* rather than journey time the basis for switching, so as to model possible national road-charging policies, and deal with elasticities to fuel price and elasticities to congestion in a consistent way; and
- improving the analysis of outputs, particularly *environmental impact* (which needs to consider distributions of speeds rather than average speeds) and economic impact of changes in traffic levels.

12.8 A National Model?

At the first seminar on National Transport Models, Henry Abraham reported on a feasibility study for a national transport model in Britain. This study was successfully concluded, and has boosted our confidence that a worthwhile national transport model is achievable, but so far has not resulted in commissioning of such a model.

This is partly due to the political emphasis currently being given to regional and local decision-making. Regional Planning Conferences (including representatives from local authorities, the regional Government Offices, and those concerned with regional economic development) are to be established and will be responsible for trunk road planning and regional transport strategies. Local authorities have been encouraged to set local traffic reduction targets, and having set targets may soon be given more powers to achieve them. Thus it may be that the efforts of the modelling profession in Britain will be concentrated on regional and local multi-modal modelling for the next few years.

The package of development work outlined above might be considered to move central government's national forecasting system a significant distance towards becoming a national transport policy model. Once the inputs are available and there is a demand for the outputs, it may in future become politically feasible to replace the current methods with a more soundly-based modelling structure.

There is an alternative possibility. It may happen that Scotland, Wales and the English regions decide to develop full regional models as part of their integrated transport and planning function, rather than local models of particular problem areas. Then, provided such models draw on consistent data sources, it may be relatively easy, in five years' time, with computer power growing ever cheaper, to link up these into some form of national model.

There's more than one way to skin a cat... *(English proverb)*

Part III

European Transport Models

13 Designing the STREAMS Model of Europe

Ian Williams
Marcial Echenique & Partners Ltd
49-51 High Street, Trumpington
Cambridge CB2 2HZ, United Kingdom

13.1 Introduction

This paper discusses some fundamental issues relating to the design and use of national or international transport models, and then briefly outlines the development of one such model, STREAMS, which models all passenger and freight movements within the EU. A detailed analysis of the factors influencing the growth in passenger travel demand is presented to illustrate some of the design principles underlying this model.

13.2 Design Alternatives

When designing national models two broad methodological strands can be distinguished:

- an empirical approach where a large budget is devoted to the collection of detailed origin-destination (OD) modal demand data on the patterns of movement throughout the country as a whole. The model is then based on making maximum use of this data directly, through expanding it into spatially detailed observed OD matrices. The UK Regional Highway Transport Model (RHTM) of the 1970s is an example of this approach. This approach is strongest when looking at short term and localised effects.
- a behavioural approach where the primary emphasis is on trying to understand the determinants of the demand for travel, and how these might change over time, space, income level, industrial or demographic segments. In this case the spatial detail is likely to be coarse, but the segmentation is likely to be detailed. This approach is strongest when looking at longer term system wide effects.

Ideally a model should be designed by starting with the behavioural approach and developing the model to a level which includes all of the main behavioural influences on travel demand in a robust and realistic fashion. Then the resulting model should be embedded within an empirical framework that allows it also to match specific local conditions and to take the observed base year travel patterns as a starting point from which the model is moved into the future. Various methods have been developed in recent years to carry out this latter task.

Sample enumeration techniques have been developed for the Dutch national transport model (Gunn, 2001) which allow responses of sampled individuals to be used to guide the changes in aggregate behaviour.

Incremental multi-level logit models of discrete choice have been developed by Bates et al. (1987) which estimate future demand changes from the observed base position as a function of the incremental change in the generalised cost of transport.

The residual disutility based method has been developed by Williams and Beardwood (1993) as a variation on the standard incremental models. A set of residual disutility values is estimated that ensures that the various choice processes within the model match a variety of sources of observed data in the base year. These residual disutility terms are then embedded in the model when executing policy test runs for future years. The estimation procedures for this method can be complex to implement initially. However, it has the advantage of producing models that are flexible enough to include a wide variety of interlocking behavioural choice mechanisms in a consistent fashion.

13.3 The STREAMS Model

In summary the STREAMS model has the following characteristics, it:

- has independent passenger and freight model components, each highly segmented;
- is a multi-modal model incorporating all modes. The model base year is 1994 and the model forecasts to 2020, transport is represented for a typical day in the year;
- has comprehensive networks for all modes and inter-modal connections; and
- includes all freight and passenger movements in the EU - it has a special treatment of local trips including walk trips.

13.3.1 Model Design Principles

It was decided at an early stage to use a zoning system based on the 'NUTS' classification (the classification used for EU data collection) to make it easier to acquire the necessary data. NUTS2 was felt to be the maximum practical level of detail, which means the model has about 200 internal zones.

The STREAMS model aims to provide medium to longer term forecasts of changes in transport demand. This implies that the model structure should be based on an understanding of why and how travel demand grows over time. The design of the model structure should also be based on assumptions about what policy options the model would be required to test and what outputs would be expected. These considerations led to a set of design principles, on the treatment of short trips, the segmentation of the model, and the approach to including transport networks.

One option in a strategic model is to include only certain types of trips, such as long distance trips. It was decided in the STREAMS project to model all trips, of whatever length, partly for technical reasons to provide a more stable basis for forecasting and partly because there would be policy interest in seeing EU transport as a 'whole'. By definition this means that in a model with NUTS2 zoning, a large proportion of trips will be within zones. This in turn implied the need to improve the treatment of intra zonal trips compared to their treatment in most transport models.

In large scale models transport networks are often represented as corridors. But the need to represent transport costs accurately, together with the level of policy interest in understanding travel patterns and especially their environmental impacts, has required the development of a full link based model.

Both the passenger and freight modules are based on the need to understand why travel demand tends to grow over time. For the passenger model this meant a detailed representation of traveller types and trip purposes. For the freight model it meant using a regional economic model to generate freight flows and segmenting these by industry type.

The emphasis in developing the model has been much closer to the behavioural than the empirical approach as defined in Section 13.2 above. In the case of passengers, there are virtually no matrices of observed longer distance trips available for use across Europe. Passenger matrices exist for all modes in Germany, for the air mode between countries (but not at the NUTS2 level), and on rail in a few countries. Car matrices do not exist in most countries. Accordingly, given current data availability, there is no possibility to adopt the empirical approach for passenger modelling at the European scale.

National passenger travel surveys do however exist for about half of the EU countries, so these do provide a solid foundation on which a behavioural model can be founded, though differences in data definitions between the different countries make direct comparisons of behaviour across countries rather complicated.

In designing the freight model the approach has been mid-way between the behavioural and the empirical. From the EUROSTAT TREX and Carriage of Goods data bases, matrices (of varying degrees of spatial detail and accuracy) of freight transport movements by mode are available for inland modes, but not for shipping. However, there are not freight equivalents of the national passenger travel surveys available, so that building up a sound behavioural description of the behaviour of freight shippers is not easy. A segmentation of freight movements by commodity type has been adopted to improve the behavioural foundations for the choice processes within the model.

13.3.2 Main Tasks

The development of the STREAMS model has involved a substantial number of detailed tasks which are summarised in Table 13.1.

Table 13.1. Main tasks in developing the STREAMS model

Demand modelling
• development of a model specification to incorporate the main mechanisms which influence passenger travel demand in terms of car availability, and behavioural and demographic change
• an in depth analysis of national passenger travel survey data was undertaken for Denmark / Finland / France / Germany / Netherlands / Sweden / UK, to develop the trip generation model
• collection/analysis of detailed zonal socio-economic/demographic data for implementing the trip generation and distribution models
• development of a regional economic model to forecast freight demand, based on national accounts and trade data
Supply modelling
• development of a comprehensive multi-modal transport modelling network
• development of cost and or tariff functions for all modes, for passengers and freight transport
• design and implementation of an innovative approach to modelling intrazonal trips, based on a distance band system
• development of a new all-day road capacity restraint function based on disaggregated data
For calibration/validation
• detailed analysis of the EUROSTAT TREX and Carriage of Goods freight databases to produce a consistent data set for freight model validation
• assemble observed passenger data across countries and modes - for calibration and validation
• model runs for base (1994) and forecast (2020) years, and also for 1975/85 as part of a 'backcasting' exercise to test the structural stability of the model

13.4 Understanding Travel Demand

When designing models to provide forecasts for a period as long as the 26 years from the base year of 1994 to the target year of 2020, it is of critical importance for the model parameters that either they will remain constant through the period or else that a well founded methodology exists to adjust them through time. This has been one of the key topics of research interest within the development of the STREAMS model.

13.4.1 Understanding the Causes of Past Growth in Passenger Traffic

Using the national travel surveys of passengers for various EU countries, an in depth study of the factors underlying the demand for travel was carried out with three main aims:

- to determine the relative magnitudes of the different factors that have influenced the demand for passenger travel and its pattern of growth over time;
- to determine the most appropriate segmentation of passenger demand into homogeneous behavioural categories; and
- to find behavioural patterns and travel statistics that either have remained constant through time and across space, or else have evolved in a predictable and stable fashion.

It is of central importance to planners to know whether road traffic growth will accelerate further in the future, or whether the rate of growth will flatten out? This is likely to be determined by the same factors that have created the observed rapid growth of demand for road space across Europe over the last 25 years. The main interrelated trends that have combined together to generate the rapid growth in vehicle kilometres on road are outlined here using either UK or EU statistics to illustrate their relative importance:

- *the population has grown through time* - this has been a minor source of road traffic growth. The UK population has grown by 4% over the 20 years to 1995 while the car kilometres have increased by 83% in this period. Likewise for the EU as a whole, a population growth of 6.5% has translated into a growth of 84% in car kilometres;
- *a greater proportion of the population are now in those population age segments that have the highest trip rates, that make the longest trips, and that are most likely to use cars*. In 1995 an annual average of 6,750 miles (DETR, 1997a, Table 2D) was travelled in the UK by those in the most mobile group, namely those between the ages of 16 and 59. This is 130% greater than the average distance for the under 15s, and 118% greater than that for the over 60s. Furthermore, this most mobile group is the population segment which also has the largest proportion of its travel taking place by car. Because demographic pyramids are not smooth through time, the last 20 years has seen relative growth in the 16 to 64 age group, compared to the growth in other age groups. From 1975 to 1994 in the EU overall it increased from 62% to 67% of the EU population (at the same time the UK went from 62% to 65%);
- *more people have access to cars,* both more households have cars, and more individuals within the household have cars available for their own use. This arises because the average household size has reduced while the average number of cars in households with two or more adults has increased rapidly.

Despite this fall in household sizes, by 1996 25% of UK households had 2 or more cars, a proportion that is more than double the rate of twenty years previously.

The increased car ownership is due largely to the average incomes of individuals increasing more rapidly than the cost of car ownership. In fact the rates of growth in the UK have been rather lower than in many other EU countries in line with the slower rate of economic growth in the UK. Higher levels of car ownership reduce the likelihood of using other modes and encourage individuals to adopt the relatively higher travel demands of the car owning segment of the population. In the UK in 1995 (DETR, 1997a, Table 2H) the annual average distance travelled across all modes was 12,500 kms per person in car owning households, but less than a third of this distance in non-car owning households, though of course the differing demographic and income characteristics will account for some of this difference in travel demand;

- *cars have captured a greater share of all the trips that are made*, because the cost-effectiveness of car travel has improved relatively more quickly than that of other modes (except air). Data in real prices from the UK (DETR,1997b, Indicator b3) show that over the 20 years from 1974, perceived marginal motoring costs (fuel plus oil) reduced by 8%, while bus and rail fares increased by 55% and 71% respectively. These public transport fares increased more rapidly than the real growth in disposable incomes of 51%. This has been a major factor in creating the growth in road travel;

- *people now tend to make longer trips due to the improved cost effectiveness of travel, especially by car.* This has been a significant source of growth since average trip lengths for car drivers have increased by 12% to 8.4 miles over the 20 years to 1995 (DETR, 1997a, Table 2.3). This growth is due both to: lower car operating costs as outlined above, and to higher average road speeds due to the greater availability of and the more intensive use of the motorways and the high quality inter-urban primary road network.

  Other modes have improved their generalised costs more slowly, if at all. However, since many trips switched into the car mode over the period, its superior performance has had an amplified effect on increasing average trip lengths;

- *people are likely to travel further due to the increased average income of individuals.* Across the EU the average journey length increases in a pronounced fashion as a function of the income of an individual. In the UK in 1995 (DETR, 1998, Table A7) the average annual distance of 11,000 miles travelled per person in a household in the highest income quintile was almost double that in the median quintile and almost three times that in the lowest quintile.

There are two main reasons for this:

- the relative damping effect of the monetary cost of travel becomes less important as incomes increase. However, as was shown above the increase in incomes was accompanied by a decrease in marginal car costs which has amplified this effect; and
- those with higher levels of disposable incomes have money to indulge their tastes for a variety of expensive activities outside the home, such as leisure, shopping and holidays. For these activities, being able to travel further allows a wider range of choice to be enjoyed;
  - *people make longer trips because of the population shift away from dense urban areas*. This increase in demand results from the migration of households and jobs to areas which are less congested for cars. It has been a major factor in generating travel growth in countries like the UK but not necessarily in all EU countries. These land use changes magnify the growth in car travel because the trips made by people in dense urban areas are likely to be shorter and are less likely to use car, than those in the areas to which population is decentralising; and
  - *car occupancy levels are lower*. The increase in car ownership has lessened the average number of occupants per car kilometre travelled, from 1.73 in 1975 to 1.62 in 1995 according to the UK NTS (DETR, 1997a, Table 2.2), so that the same number of person journeys by car in 1995 would now generate 7% more vehicles to make these journeys than in 1975.

It is the combination of how each of the above eight mechanisms evolves that determines the rate at which the future demand for transport will grow in an area. These mechanisms in general are not independent of one another but are highly correlated. This implies that considerable sophistication is required in the model structure in order to co-ordinate the multidimensional nature of these influences, and thus produce coherent, consistent forecasts of travel demand. All of these mechanisms, except the item relating to population relocation, are represented explicitly in the STREAMS model.

13.4.2 Broad Trends in Freight and Passenger Travel Demand

The list above of eight significant factors that significantly influence the evolution of the demand for transport through time, deliberately does not include increases through time in the number of trips per person. Once the population has been segmented into suitably homogeneous groups, it appears from empirical observation that overall trip rates per person have increased only slowly through time. Leisure and holiday trips have increased significantly, but these comprise only a small proportion of trips overall. This relative stability in trip rates is true only if all trips (i.e. including short trips by modes walk and cycle) are included. It is not true if mechanised trips are looked at in isolation.

Fig. 13.1 shows the growth through time in the UK in the yearly number of trips, of travel hours and of passenger kilometres per person on all forms of passenger transport. Here the travel distance has increased rapidly, but the travel time and number of trips per person have both grown quite slowly and at similar rates. This confirms many of the trends discussed in the pervious section.

The average number of trips made by a particular type of person exhibits a reasonable similarity across the EU. Obviously when broken down by trip purpose, different categories of people have different needs to travel. However, taking all trips together, in general the young and the retired make the fewest trips while those in employment make the most trips. The number of trips made does not vary enormously either over the years, or between countries. Accordingly, we should not expect trip rates per person to increase rapidly into the future.

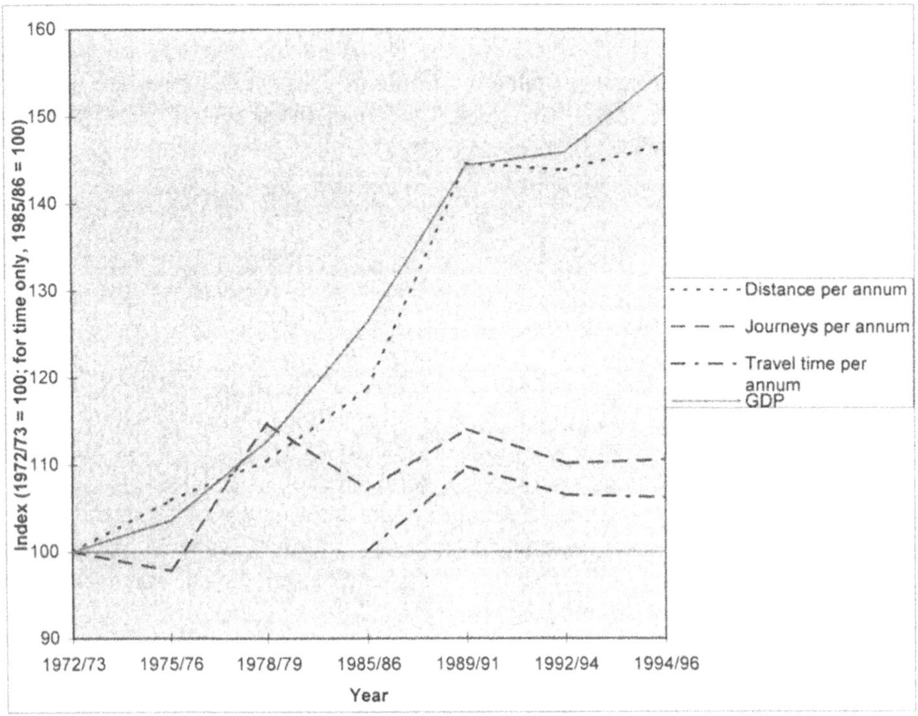

Fig. 13.1. Annual number of journeys, kilometres travelled, and travel time per person and GDP in the UK (data on travel time are not available prior to 1985/86).
Source: UK National Travel Survey (DETR, 1997a)

Turning now to freight, the growth of freight carried on heavy goods vehicles on roads in the UK is presented in various ways in Fig. 13.2. The number of tonnes moved by road has increased sporadically by only 12% overall in the 22 years to 1996 in the UK, while the average length of haul has increased over the same period by 50%. As recently as 1992, the tonnes lifted were almost the same as in

1974 which suggests little growth, while the growth in tonne kilometres has been substantial at 69% over this 22 year period.

Although these figures presented in Fig. 13.2 are for roads in the UK, the message is the same for road freight for EU countries in general (DG VII, 1998, Table 4.4). From 1970 to 1996 the road tonne kilometres in the EU increased by 167% while the road tonnes lifted increased at a much slower rate. On other inland freight modes a similar, though less pronounced trend also occurred.

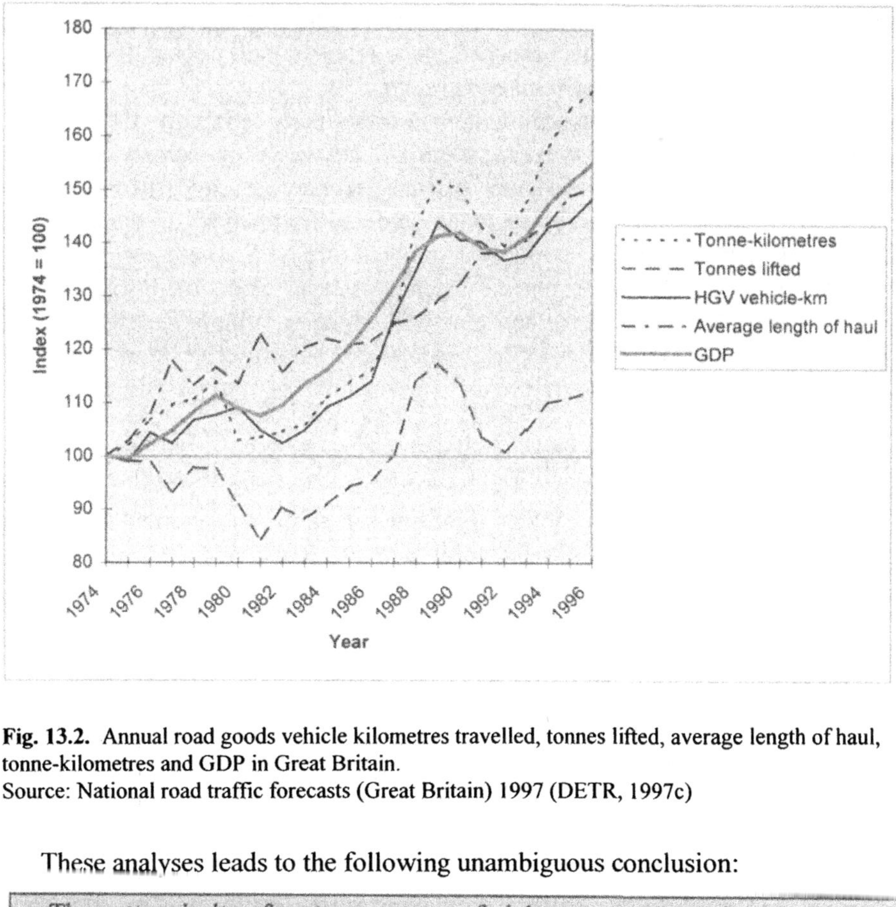

Fig. 13.2. Annual road goods vehicle kilometres travelled, tonnes lifted, average length of haul, tonne-kilometres and GDP in Great Britain.
Source: National road traffic forecasts (Great Britain) 1997 (DETR, 1997c)

These analyses leads to the following unambiguous conclusion:

The vast majority of past passenger or freight transport growth has arisen from increases in the *length* of journeys, rather than increases in the *number* of trips per person or of tonnes carried.

Accordingly, models that operate primarily in terms of changes in the numbers of tonnes or of trips carried, rather than changes in the distances travelled, are unlikely to provide well founded forecasts of future transport patterns or of responses to policy initiatives.

13.5 Uses of the STREAMS Model

The strength of the STREAMS model is in producing EU wide transport forecasts of the impacts of socio-economic changes and transport policy changes, together with an indication of the distribution of results by country and region. The model design allows countries with different behavioural and socio-economic characteristics to be modelled successfully within a consistent overall structure. The detailed multi-modal approach coupled with a precise representation of costs and tariffs allows a wide range of policies to be tested including the impact of differential price changes between the modes. The highly segmented passenger and freight demand models provide a mechanism to represent the impact of economic and demographic change on the transport system.

However, there are inevitably limitations on the likely reliability of the model results at a detailed spatial level. A model of the whole of Europe cannot be expected to produce the same degree of local accuracy as, say, fifteen national models, or a greater number of regional models. The nature of these spatial limitations is being explored in a follow up project, SCENES. There also remains a general technical issue of the extent to which a strategic model of this nature can provide accurate flows across the entire network. Incorporating a real network, as opposed to a stylised or corridor representation does help to achieve accurate cost and time estimates, but obviously the link flow results themselves will also be of interest.

The STREAMS model is currently being used as part of an EC Pilot Strategic Environmental Assessment of the Trans-European Networks (TENs) to measure the magnitude and the broad locations of the environmental changes that arise from the development of the TENs and from any accompanying transport policy initiatives.

Acknowledgements

The creation of the STREAMS model has been financed as a Fourth Framework Research project by DG VII of the European Commission. The STREAMS research team comprises: IRPUD, Germany; LT Consultants, Finland; MECSA, Spain; ME&P, UK; and TRT, Italy. However, the views expressed here are those of the author alone.

References

Bates, J., Ashley, D. and Hyman, G. (1987), 'The nested incremental logit model: theory and application to modal choice', in *Proceedings of the PTRC Summer Annual Conference, Series C*, PTRC Education and Research Services, London, pp. 75-82.

Gunn, H. (2001), 'An overview of European national models', Chapter 2, this volume.

DETR (1997a), *National Travel Survey 1994/96*, The Stationery Office, London.

DETR (1997b), *Indicators of Sustainable Development for the United Kingdom*, The Stationery Office, London.

DETR (1997c), *National Road Traffic Forecasts (Great Britain) 1997*, Department of the Environment, Transport and the Regions, London.

DETR (1998), *Transport Trends*, The Stationery Office, London.

DG VII (1998), *EU Transport in Figures*, Office for Official Publications of the European Community, Luxemburg.

Williams, I.N. and Beardwood, J.E. (1993), 'A residual disutility based approach to incremental transport models', in *Proceedings of the PTRC Summer Annual Conference, Series D*, PTRC Education and Research Services, London, pp. 11-22.

14 Tests of Nonlinearity, Modal Captivity and Spatial Competition within the STEMM Multicountry Application for Passengers

Marc Gaudry
Bureau d'économie théorique et appliquée (BETA)
Université Louis Pasteur, Strasbourg, France
and
Centre de recherche sur les transports (CRT)
Université de Montréal, Montréal, Canada

14.1 Introduction: The STEMM Passenger Demand Modelling Project

The passenger demand stream of the STrategic European Multimodal Modelling (STEMM) effort attempted to establish that a multicountry model of use in the study of multimodality and of intermodality could be built and that advanced techniques of demand analysis, going beyond currently applied state-of-the-art nonlinear techniques, could be fruitfully applied to it.

Our present summary will neglect all work done on the construction of multimodal networks (Eberhard et al., 1998) and focus on the demand models proper, drawing heavily from the full report on demand methodology (Gaudry et al., 1998). We summarise the baseline model, originally developed for Germany and Canada within the Quasi-Direct Format (QDF) framework. We describe the multicountry application MAP-1 effected with British and French databases and the advanced tests performed on the MAP-1 business trip submodel.

14.2 The Baseline Framework and Model

14.2.1 The Quasi-Direct Framework Components

The baseline approach consists in a two-step three-component model format, called the Quasi-Direct Format, that explains demand by mode T_{ijm} as a product of two coupled models where the first explains total demand for all modes T_{ij} but includes a measure of the utility of all modes, U_{ij}, and the second splits this demand with a mode choice model. Coupling arises from the denominator of that model used as overall utility measure:

$$T_{ijm} = T_{ij} \cdot p_{ijm} \tag{1}$$

or, more explicitly:

$$T_{ijm} = f(A_i, A_j, U_{ij}) \cdot \left(U_{ijm} \Big/ \sum_m U_{ijm} \right) \qquad m = 1, \ldots, M \tag{2}$$

with

$$U_{ij} = \sum_m U_{ijm} \tag{3}$$

In the baseline model (Gaudry et al., 1994), the first model is an aggregate flow Generation-Distribution model and the mode choice model is an aggregate Box-Cox logit model explaining shares of the intercity modes (3 for Germany and 4 for Canada). The latter can be expressed, dropping the origin-destination subscripts and using *i* to denote the mode, as:

$$U_i = exp(V_i) \tag{4}$$

$$V_i = \beta_{i0} + \sum_n \beta^i_{in} X_n^{i(\lambda^i_{in})} + \sum_s \beta_{is} X_s^{(\lambda_{is})} \tag{5}$$

where the Box-Cox (BC) transformation is applied. Direct and inverse Box-Cox and Box-Tukey forms, defined with parameters λ_K and μ_K on the positive variable X_k, are defined in Table 14.1. They are all of relevance to STEMM.

Table 14.1. Transformations defined by Box and Cox (1964) and Gaudry (1981)

Box and Cox (1964)		Gaudry (1981)	
Direct Power Transformation (DPT)		Inverse Power Transformation (IPT)	
$X_k^{(\lambda_k, \mu_k)}$	Conditions	$Z_k^{(\phi_k, \mu_k)^{-1}}$	Conditions
BC		BCG	
$\frac{X_k^{\lambda_k} - 1}{\lambda_k}$	$\lambda_k \neq 0, X_k \rangle 0$	$(\phi_k Z_k + 1)^{1/\phi_k}$	$\phi_k \neq 0, (\phi_k Z_k + 1) \rangle 0)$
$ln X_k$	$\lambda_k = 0, X_k \rangle 0$	$exp(Z_k)$	$\phi_k = 0, (Z_k \rangle 0)$
BT		BTG	
$\frac{(X_k + \mu_k)^{\lambda_k} - 1}{\lambda_k}$	$\lambda_k \neq 0, (X_k + \mu_k) \rangle 0)$	$(\phi_k Z_k + 1)^{1/\phi_k} - \mu_k$	$\phi_k \neq 0, (\phi_k Z_k + 1) \rangle 0)$
$ln(X_k + \mu_k)$	$\lambda_k = 0, (X_k + \mu_k) \rangle 0)$	$exp(Z_k) - \mu_k$	$\phi_k = 0, (Z_k \rangle 0)$

14.2.2 The Issues of Form and Spatial Competition

In both logit and generation-distribution models, the Box-Cox form has long demonstrated its ability to give better fits and more reasonable results than *ad hoc* forms (notably linear), to influence the elasticities, values of time and even the signs of regressors (Gaudry and Wills, 1978)!

It was therefore a necessary dimension of the STEMM models. But in models of levels, such as the generation-distribution model, the baseline model had used a formulation that, in addition to the question of form, simultaneously addressed that of heteroskedasticity and spatial correlation of the residuals with the following 4-equation specification:

$$y_t^{(\lambda_y)} = \sum_{k=1}^{K} \beta_k \cdot X_{kt}^{(\lambda_{x_k})} + u_t \tag{6}$$

$$u_t = \left[\exp\left(\sum_m \delta_m Z_{mt}^{(\lambda_{z_m})} \right) \right]^{1/2} \cdot v_t \tag{7}$$

$$v_t = \sum_{l=1}^{2} \rho_l \left(\sum_{n=1}^{N} \tilde{r}_{l,tn} v_n \right) + w_t \tag{8-A}$$

where a positive rho (ρ) coefficient denotes competition and a negative one complementarity and $\tilde{r}_{l,tn}$ is the typical element of the matrix $\tilde{R}_l$. In matrix form:

$$v = \sum_{l=1}^{2} \rho_l \tilde{R}_l v + w \tag{8-B}$$

$$\tilde{R}_l = \pi_l \left[I - (1-\pi_l)\overline{R}_l \right]^{-1} \overline{R}_l \qquad (0 \prec \pi_l \leq 1) \tag{9}$$

where the proximity parameter π_l allows endogenisation of the relative importance of near and distant effects embodied in the Boolean neighbourhood matrix. If π_l equals one, $\tilde{R}_l$ is equal to $\overline{R}_l$, indicating that only the adjacent neighbours have an impact on the correlations among the associated residuals. By contrast, as π_l tends towards zero, the near effect is reduced to a minimum in favour of the distant effect.

Generally, the π_l weigh the *relative* importance of near and distant effects with a single parameter defining the sharpness or slope of the decline. The baseline cases had used relatively few origin-destination pairs to apply this procedure and issues of size of matrix R were part and parcel of STEMM advanced tests.

14.3 A First Multicountry Application for Passengers

14.3.1 A Multicountry Strategy

Data Foundations and Simple Specifications at a NUTS-3 Level. In the absence of Europe-wide passenger travel surveys, the development of a multicountry model on the NUTS-3 level defined by EUROSTAT (1992) required a specification that meets the lowest common data availability denominator across national databases that contain cross-country trips. Despite considerable, and sometimes vain, efforts – that still provided us with national data sets for various countries (Norway, Sweden, Germany) – we were limited in the availability of data sets that contained significant cross-border trips: this is not a recent problem (COST 305, 1988). The best such recent bases are the French 'Enquête Transports 1993-1994' (INSEE, 1993), the British Office for National Statistics (ONS) Channel Tunnel related International Passenger Surveys of 1991 and 1996 and the Heathrow-related Civil Aviation Authority (CAA) data sets. Given more resources and time, this tripod could have been marginally expanded with other, more limited and partial sources. In the tables below, "SES" denotes French data and "CAA/IPS" British data.

For intercity travel demand modelling purposes, the NUTS-3 level ('Départements' in France, 'Kreise' in Germany and Counties in The United Kingdom, etc.) constrains the availability of socio-economic variables for our two model stages. This means that the mode choice model cannot easily contain many socio-economic variables of the kind found in recent discrete choice models. Fortunately, a model based on network variables can be specified: it will resemble the early two-stage models such as McLynn et al. (1968) and McLynn and Woronka (1969). To the extent that there has been a drift over time of variables into the mode choice model from other stages of the decision process, and that this drift has not been determined on statistical grounds by procedures that would jointly determine the role of all variables in each stage, a simple specification in terms of modal cost and time characteristics should be acceptable. Similarly, the Generation-Distribution model will be restricted in its potential list of factors but, as it is notoriously difficult to do better than a population based gravity model at any time, this limitation is relatively inconsequential. Overall, such a structure should produce a robust model.

The Resulting MAP-1 Model. The Level-Probability combination made possible by available survey data provided an opportunity to make the best use of the relative performance of aggregate and disaggregate data. So two series of models were studied with national data, primarily for Germany (*Mobilität 1991* data) but also with other national databases, to study the robustness of the intended format and specification. Each stream of work aimed at doing at least as well, for each of 3 trip purposes, as the baseline model had done without a distinction by trip purpose, using the TRIO program (Gaudry, 1997).

Validated approaches were then applied to multicountry data sets. Selected results for model variants produced by the Tablex utility of TRIO (see Tables 14.2-14.5) include, in three sections, subsets of:

- in Part I, coefficients, partial derivatives, elasticities with corresponding t-statistics of the underlying β_k or δ_Z, as well as other indicators (genericity of β_k; presence of a λ_k). Code names underlined once denote quasi-dummy variables and those underlined twice denote true dummy variables;
- in Part II, coefficients of form or captivity (BC, BT, BCG and BTG), of heteroskedasticity and directed autocorrelation, some shown only with their usual t-statistic against 0, but others, like the proximity parameter π, shown also with the t-statistic against 1; and
- in Part III, general statistics, especially the sample size and the log-likelihood value.

Mode choice: asymmetry of response. The first stream of work resulted in the generic mode choice models based on the notion of principal mode, shown in Table 14.2. Genericity means that the weights associated to the network variables are constrained equal, a procedure that, in the general literature, is often deemed necessary to obtain reasonable signs. Although all definitions of variables in Table 14.2 are self-explanatory, it should be remembered that out-of-vehicle time is a component of total travel time and therefore captures the *supplementary* disutility of this kind of time. There is no dummy variable to account for heterogeneity among data sources. Note that about half of the huge sample of about 78,000 observations pertains to vacation trips.

For each trip purpose, the first column (MAP 1.1, MAP 2.1 and MAP 3.1) of Table 14.2 shows results for the usual linear model and the second column (MAP 1.2, MAP 2.2 and MAP 3.2) shows the impact of a single Box-Cox transformation. In the three models, this BC induces astronomical gains in log-likelihood value, more significant explanatory variables and implies an asymmetric response function, the first such demonstration for a multi-country model.

Generation-Distribution: rejection of the multiplicative homoskedastic form. The second stream of work improves the standard formulation of distribution models in three ways: (i) the population term at the origin is weighted by the propensity to travel by age class, an idea first developed in Last (1998) and applied here (with fixed weights derived from the French data); (ii) the impedances include both language and the presence of a border, in addition to the utility term from a Logit model; (iii) interesting attractors are defined for vacation trips. The sample of 21,000 Origin-Destination flows contains 40% of private purpose trips and is strongly home-based in France and the United Kingdom.

Box-Cox transformations (as in (6)) and heteroskedasticity (of the general type (7) – and very convincingly based on the Z variable distance) were introduced. Table 14.3 summarises the work carried out by focusing on the contrast between the usual multiplicative gravity-like homoskedastic specification for each trip purpose (MAP 1.1, MAP 2.1 and MAP 3.1) and the more general BC-HE (Box-Cox heteroskedastic) formulation (MAP 1.2, MAP 2.2 and MAP 3.2). The

Table 14.2. MAP-1 reference asymmetric response mode choice models (estimated by F. Heinitz). Logit family. Standard linear and standard Box-Cox types (PROBABILITY P-2 algorithm). Choice probability for three modes; business, private and vacation trips (CAA/IPS 1996 data)

I. W.A. ELASTICITIES and (COND. T-STATISTIC)							
	VARIANT	=MAP 1.1	MAP 1.2	MAP 2.1	MAP 2.2	MAP 3.1	MAP 3.2
	FAMILY	=LOGIT	LOGIT	LOGIT	LOGIT	LOGIT	LOGIT
	TYPE	=linlogit	bcllogit	linlogit	bcllogit	linlogit	bcllogit
	PURPOSE	=BUSINESS	BUSINESS	PRIVATE	PRIVATE	VACATION	VACATION
ALTERNATIVE 1 : AIR							
NETWORK VARIABLES							
TOTAL GROUP COST [AIR]	TXCostA	-.07	-.09	-.05	-.09	-.09	-.10
		(-18.61)	(-20.30)	(-10.01)	(-12.74)	(-23.78)	(-24.32)
	(1)	L 1(GE)	L 1(GE)	L 1(GE)	L 1(GE)	L 1(GE)	L 1(GE)
TOTAL TRAVEL TIME [AIR]	TTTimeA	-.05	-.14	-.05	-.14		
		(-14.25)	(-23.03)	(-9.47)	(-17.77)		
	(1)	L 1(GE)	L 1(GE)	L 1(GE)	L 1(GE)		
TOTAL OUT-OF-VEHICLE [AIR]	TOoVTimA	-.02	-.03	-.01	-.03		
		(-2.35)	(-6.59)	(-1.44)	(-2.98)		
	(1)	(GE)	(GE)	(GE)	(GE)		
REGRESSION CONSTANT	CONSTANT	-	-	-	-	-	-
		(34.31)	(31.05)	(35.53)	(35.38)	(-16.03)	(-41.14)
	(1)						
ALTERNATIVE 2 : RAIL							
NETWORK VARIABLES							
TOTAL GROUP COST [RAIL]	TXCostR	-.57	-1.40	-.15	-.88	-.20	-.20
		(-18.61)	(-20.30)	(-10.01)	(-12.74)	(-23.78)	(-24.32)
	(2)	L 1(GE)	L 1(GE)	L 1(GE)	L 1(GE)	L 1(GE)	L 1(GE)
TOTAL TRAVEL TIME [RAIL]	TTTimeR	-1.00	-1.32	-.54	-.91	-1.55	-1.63
		(-14.25)	(-23.03)	(-9.47)	(-17.77)	(-68.32)	(-70.66)
	(2)	L 1(GE)	L 1(GE)	L 1(GE)	L 1(GE)	L 1(GE)	L 1(GE)
TOTAL OUT-OF-VEHICLE [RAIL]	TOoVTimR	-.09	-.21	-.04	-.09		
	--------	(-2.35)	(-6.59)	(-1.44)	(-2.98)		
	(2)	(GE)	(GE)	(GE)	(GE)		
TOTAL NUMBER OF TRANSFERS [RAIL]	TUmstR	-1.11	-1.06	-2.04	-1.73	-.17	-.18
		(-14.25)	(-15.93)	(-24.68)	(-22.00)	(-5.10)	(-5.29)
	(2)	(SP)	(SP)	(SP)	(SP)	(SP)	(SP)
REGRESSION CONSTANT	CONSTANT	-	-	-	-	-	-
		(19.09)	(18.43)	(26.97)	(19.45)	(7.17)	(7.55)
	(2)						

Table 14.2. (continued)

I.W.A. ELASTICITIES and (COND. T-STATISTIC)							
	VARIANT	=MAP 1.1	MAP 1.2	MAP 2.1	MAP 2.2	MAP 3.1	MAP 3.2
	FAMILY	=LOGIT	LOGIT	LOGIT	LOGIT	LOGIT	LOGIT
	TYPE	=linlogit	bcllogit	linlogit	bcllogit	linlogit	bcllogit
	PURPOSE	=BUSINESS	BUSINESS	PRIVATE	PRIVATE	HOLIDAY	HOLIDAY
ALTERNATIVE 4 : CAR							
NETWORK VARIABLES							
TOTAL COST [CAR]	TCostC	-.76	-1.25	-.22	-.81	-.22	-.22
		(-18.61)	(-20.30)	(-10.01)	(-12.74)	(-23.78)	(-24.32)
	(4)	L 1(GE)	L 1(GE)	L 1(GE)	L 1(GE)	L 1(GE)	L 1(GE)
TOTAL TRAVEL TIME [CAR]	TTTimeC	-1.12	-1.34	-.72	-.91	-1.60	-1.66
		(-14.25)	(-23.03)	(-9.47)	(-17.77)	(-68.32)	(-70.66)
	(4)	L 1(GE)	L 1(GE)	L 1(GE)	L 1(GE)	L 1(GE)	L 1(GE)
TOTAL OUT-OF-VEHICLE [CAR]	TOoVTimC	-0...	-0...	-0...	-0...		
	--------	(-2.35)	(-6.59)	(-1.44)	(-2.98)		
	(4)		(GE)	(GE)	(GE)	(GE)	

II. PARAMETERS UNCOND.[T-STATISTIC=0] UNCOND.[T-STATISTIC=1]							
BOX-COX TRANSFORMATIONS							
LAMBDA(X) - GROUP 1	LAM 1	1.00	-.62	1.00	-.33	1.00	-.53
		FIXED	[-13.39]	FIXED	[-7.47]	FIXED	[10.79]
			[-34.99]		[-29.84]		[-9.51]

III.GENERAL STAT.							
	VARIANT	=MAP 1.1	MAP 1.2	MAP 2.1	MAP 2.2	MAP 3.1	MAP 3.2
	FAMILY	=LOGIT	LOGIT	LOGIT	LOGIT	LOGIT	LOGIT
	VARIANT	=linlogit	bcllogit	linlogit	bcllogit	linlogit	bcllogit
	PURPOSE	=BUSINESS	BUSINESS	PRIVATE	PRIVATE	HOLIDAY	HOLIDAY
LOG-LIK.-FINAL VALUE		-6731.15	-6246.86	-7288.48	-7117.35	-20343.0	-20299.7
-INITIAL VALUE		-7286.77	-24563.9	-7926.74	-7926.74	-25422.2	-25422.2
-WITH CONSTANTS ONLY		-7286.77	-7286.77	-7926.74	-7926.74	-25422.2	-25422.2
-RATIO TEST		1111.24	2079.82	1276.52	1618.77	10158.46	10245.05
RHO-SQUARED		.08	.14	.08	.10	.20	.20
RHO-SQUARED BAR -AKAIKE		.08	.14	.08	.10	.20	.20
-HOROWITZ		.08	.14	.08	.10	.20	.20
-HENSHER & JOHNSON		.08	.14	.08	.10	.20	.20
PER CENT RIGHT		92.07	92.18	87.67	87.61	76.24	77.50
SAMPLE -NUMBER OF ALTERNATIVES		3	3	3	3	3	3
-NUMBER OF OBSERVATIONS		22579	22579	17477	17477	37512	37512
-FIRST OBSERVATION		1	1	1	1	1	1
-LAST OBSERVATION		22579	22579	17477	17477	37512	37512
-AVAILABLE OBSERVATIONS		22359	22359	17308	17308	35004	35004
-AVAIL OBS IN 1 AIR		22579	22579	17477	17477	37512	37512
-AVAIL OBS IN 2 RAIL		22579	22579	17477	17477	37512	37512
-AVAIL OBS IN 4 CAR		22579	22579	17477	17477	37512	37512
NUMBER OF ESTIMATED PARAMETERS							
-BETAS .VARIABLES		4	4	4	4	3	3
.CONSTANTS		2	2	2	2	2	2
.ASSOCIATED DUMMIES		0	0	0	0	0	0
-BOX-COX TRANSFORMATIONS		0	1	0	1	0	1

Note: under each t-statistic found in Section I, the symbols denote a generic (GE) or specific (SP) regression coefficient, as well as (L), the presence of a BCT and its group identification.

Table 14.3. MAP-1 reference flexible form generation-distribution models (estimated by J. Last). BC-GAUHESEQ family (Level-1.4 algorithm). Multiplicative Box-Cox (2-3) and heteroskedasticity types. Total flow by all modes. Business, private and vacation trips (CAA/IPS 1991 and SES 1993-1994 data)

I.ELASTICITY and (COND. T-STATISTIC) E(Y)	VARIANT = FAMILY = TYPE = PURPOSE =	MAP 1.1 LEVEL-1 LOG BUSINESS	MAP 1.2 LEVEL-1 BC3+HE BUSINESS	MAP 2.1 LEVEL-1 LOG PRIVATE	MAP 2.2 LEVEL-1 BC2+HE PRIVATE	MAP 3.1 LEVEL-1 LOG VACATION	MAP 3.2 LEVEL-1 BC3+HE VACATION
ORIGIN CHARACTERISTICS							
WORKING POP. WEIGHTED AT ORIGIN	wpopxpsi3	.32 (14.13) LAM 1	.33 (17.65) LAM 1				
POPULATION WEIGHTED AT ORIGIN	popxpsi3			.34 (19.35) LAM 1	.34 (21.32) LAM 1	.27 (23.05) LAM 1	.26 (24.09) LAM 1
GDP PER WORKER AT ORIGIN	o_gdppwc	.55 (8.11) LAM 3	.55 (9.43) LAM 3				
GDP PER CAPITA AT ORIGIN	o_gdppc			-.01 (-.21)	.05 (1.66)	.06 (2.32)	.10 (3.99)
AREA OF ORIGIN	o_area	-.04 (-2.06) LAM 1	-.06 (-3.23) LAM 1				
DESTINATION CHARACTERISTICS							
WORKING POP. AT DESTINATION	d_workpop	.43 (18.29) LAM 1	.44 (23.71) LAM 1				
POPULATION AT DESTINATION	d_pop			.50 (24.90) LAM 1	.50 (28.16) LAM 1	.14 (8.65) LAM 1	.13 (8.47) LAM 1
GDP PER WORKER AT DESTINATION	d_gdppwc	.50 (8.02) LAM 3	.47 (9.27) LAM 3				
AREA OF DESTINATION	d_area	-.08 (-4.15) LAM 1	-.10 (-5.69) LAM 1				
SEASIDE AT DESTINATION	seaside =======			.12 (3.52)	.13 (4.06)	.15 (5.39)	.15 (5.93)
MOUNTAINS AT DESTINATION	mountain ========					.30 (6.69)	.32 (7.02)
JULY TEMPERATURE DESTINATION/ORIGIN RATIO	summer2					1.07 (16.24) LAM	1.27 (17.52) LAM
IMPEDANCES							
UTILITY	util_int3	.06 (8.01) LAM	.23 (12.50) LAM	.33 (20.42) LAM	.54 (27.21) LAM	.18 (8.91) LAM	.24 (13.29) LAM
BORDER IS CROSSED	cross =====	-.82 (-7.21)	-.81 (-6.46)	-.46 (-4.13)	-.39 (-3.68)	-.33 (-5.27)	-.26 (-4.18)
COMMON LANGUAGE ACROSS BORDER	lang2 =====			.30 (1.43)	.13 (.67)	.57 (3.13)	.50 (3.38)
DATABASE							
CHANNEL SURVEY	ips_caa =======	-2.61 (-22.73)	-2.01 (-15.24)	-2.47 (-20.50)	-1.99 (-17.04)	-1.79 (-29.55)	-1.75 (-28.98)

Table 14.3. (continued)

RECORD # 1642	rec1642			3.09	3.16		
OUTLIER	=======			(2.30)	(.00)		
RECORD # 1912	rec1912					2.70	2.61
OUTLIER	=======					(2.86)	(.00)
ASSOCIATED DUMMIES GROUP							
WORKING POP.	wpopxpsi3	-3.46	-2.46				
WEIGHTED AT ORIGIN	=========	(-5.57)	(-2.49)				
REGRESSION CONSTANT .	CONSTANT	-	-	-	-	-	-
		(12.21)	(8.88)	(2.00)	(17.02)	(18.97)	(21.36)
HETEROSKEDASTICITY STRUCTURE							
DELTA COEFFICIENTS							
DISTANCE BY CAR	dist		-.17		.02		.01
			(-10.39)		(4.57)		(10.04)
			LAM		LAM		LAM

II. PARAMETERS and (COND. T-STATISTIC)	VARIANT / TYPE / PURPOSE	=MAP 1.1 =LOG =BUSINESS	MAP 1.2 BC3+HE BUSINESS	MAP 2.1 LOG PRIVATE	MAP 2.2 BC2+HE PRIVATE	MAP 3.1 LOG VACATION	MAP 3.2 BC3+HE VACATION
HETEROSKEDASTICITY STRUCTURE							
BOX-COX TRANSFORMATIONS: UNCOND: [T-STATISTIC=0] / [T-STATISTIC=1]							
LAMBDA(Z)	dist		.78		3.82		5.56
			[2.47]		[2.18]		[7.42]
			[-.68]		[1.61]		[6.08]
BOX-COX TRANSFORMATIONS: UNCOND: [T-STATISTIC=0] / [T-STATISTIC=1]							
LAMBDA(Y) - GROUP 1	LAM 1	.00	.05	.00	.03	.00	.01
		FIXED	[7.06]	FIXED	[5.82]	FIXED	[1.26]
			[-129.3]		[-212.1]		[-156.8]
LAMBDA(X)	util_int3	.00	.33	.00	.31	.00	0...
		FIXED	[9.56]	FIXED	[14.61]	FIXED	[.06]
			[-19.20]		[-33.35]		[-15.38]
LAMBDA(X)	summer2					.00	1.02
						FIXED	[1.67]
							[.03]
LAMBDA(X) - GROUP 1	LAM 1	.00	.05	.00	.03	.00	.01
		FIXED	[7.06]	FIXED	[5.82]	FIXED	[1.26]
			[-129.3]		[-212.1]		[-156.8]
LAMBDA(X) - GROUP 3	LAM 3	.00	-.06				
		FIXED	[-.33]				
			[-5.41]				

III.GENERAL STATISTICS	VARIANT= / TYPE =	MAP 1.1 LOG	MAP 1.2 BC3+HE	MAP 2.1 LOG	MAP 2.2 BC2+HE	MAP 3.1 LOG	MAP 3.2 BC3+HE
LOG-LIKELIHOOD		-50852.7	-50737.6	-80125.0	-79986.1	-65405.3	-65327.1
SAMPLE : - NUMBER OF OBSERVATIONS		5711	5711	8334	8334	6540	6540
- FIRST OBSERVATION		1	1	1	1	1	1
- LAST OBSERVATION		5711	5711	8334	8334	6540	6540
NUMBER OF ESTIMATED PARAMETERS :							
- FIXED PART :							
. BETAS		11	11	10	10	12	12
. BOX-COX		0	3	0	2	0	3
. ASSOCIATED DUMMIES		1	1	0	0	0	0
- AUTOCORRELATION		0	0	0	0	0	0
- HETEROSKEDASTICITY :							
. DELTAS		0	1	0	1	0	1
. BOX-COX		0	1	0	1	0	1
. ASSOCIATED DUMMIES		0	0	0	0	0	0

multiplicative form, including the *famous log-sum aggregator (log of the utility variable), is rejected* in all models despite the fact that the actual values of the BC are numerically quite close to zero.

There may perhaps be an exception to this general conclusion in the vacation model (MAP 3.2) where low *t*-statistics with respect to zero appear to contradict the likelihood ratio tests results. In this case, it is possible that the approximate method (Berndt et al., 1974) used to compute *t*-statistics from first derivatives of the log-likelihood is too imprecise when the true value of the coefficient is close to zero – in which case the *t*-test lacks power anyway – and that one should not go beyond the exact likelihood ratio test for this model. This minor point need not detract from the overall value of this powerful example: the generalisation of the multiplicative form would be impressive in a national model and is of no less interest in this first-ever context of a multinational model.

Detailed examination of the results found in Table 14.3 are also interesting:

- the sums of population and GDP per person elasticities is less than 1, and differs across purposes;
- although all elasticities of the utility term increase with the BC, private trips are always and remain the most sensitive, twice as high (at 0,54) as the other trip purposes (at 0,23 and 0,24);
- border effects are larger for business than for other purposes and smallest for vacation trips, but language effects do not exist for business trips and are strongest for vacation trips; and
- Channel survey flows are much lower than other flows, reflecting perhaps both sampling design and geographical barrier effects.

14.3.2 Tests of Algorithmic Developments under STEMM

Although these MAP-1 results constitute first instances of multi-country applications of the modal asymmetry and generalized gravity specifications, they are not strictly speaking tests of the most advanced program design improvements carried out under STEMM. In models of levels, these design improvements were concentrated on the addition of directed autocorrelation (notably spatial) options usable with large samples of an aggregate nature; in choice models, the design improvements featured generalizations of the logit model to allow captivity/ignorance (thick tails) and overall asymmetry of the response curve (of the V_i representative utility function or of the U_i attraction function) with samples of a discrete nature.

To demonstrate the usefulness of the new designs, we performed additional tests, but only on the business trip models. To ease the comparison of results, we produced for the mode choice and generation-distribution models new tables where MAP-1 results already listed in Tables 14.2-14.3 are reproduced along with the more general results to be discussed presently.

Demonstration of New Dimensions: Mode Choice. In Table 14.4, the first two columns reproduce the MAP-1.1 and MAP-1.2 models of the form:

$$U_i = \exp(V_i), \quad V_i = \beta_{i0}^{\nabla} + \sum_n \beta_{in}^{i} X_n^{i(\lambda_n)} \tag{10}$$

and we note in passing that the optimal form for the X_n is $1/X^{0,6}$, almost one over square root of the network variable. The existence of a negative λ on a network variable often indicates that the *expenditure* specification of network variables (fare, time) might be changed to the *rate* specification (price, speed, distance) in order to best capture the implicit money and time income effects that may occur in demand for long-distance trips: these specifications are only equivalent in logarithmic form.

We wish to test whether such estimates leave much to be still explained (captivity/ignorance) and properly account for the form of the response curve despite the clear presence of non-linearity on the variables. We therefore use the Box-Tukey Inverse Power Transformation - Logit (BT-IPT-Logit) family with a representative utility function of Standard Box-Cox (S-BC) type. We choose the BT-IPT instead of the (linear) LIN-IPT family defined as:

$$U_i = \left[\varphi_i \, exp(V_i) + 1\right]^{1/\varphi_i} - \mu_i \tag{11}$$

because the latter tends to require more computing time than the former. We continue with the S-BC function from (10), without adding a third constant, to insure simple continuity with the logit family results: the third constant would require an intermediate model and make the models non-nested because a formulation with M-1 identified alternative-specific constants is not nested in a formulation with M identified alternative-specific constants. Our demonstration model is therefore:

$$U_i = \exp\left\{\frac{\left[\exp(V_i) + \tilde{\mu}_i\right]^{\tilde{\varphi}_i} - 1}{\tilde{\varphi}_i}\right\}, \quad V_i = \beta_{i0}^{\nabla} + \sum_n \beta_{in}^{i} X_n^{i(\lambda_n)} \tag{12}$$

and we propose to analyze four different variants, called D-1.3 to D-1.6 in Table 14.4. Due to the fact that M-1 alternative-specific constants are used (and not M), the results shown *would not be invariant* to eventual changes in the units of measurement of the X_n , but this is immaterial for our purposes, as the reader can infer from the great stability of the results (elasticities, BC transformations), because units are not modified as we pass from logit columns (where they would be invariant) to the IPT columns.

Captivity/ignorance. First, we start from the logit case and introduce a common $\tilde{\mu}$ on all three alternatives. The great increase of more than 13 points in log-likelihood from column MAP 1.2 to column D 1.3 shows that captivity/ignorance

Table 14.4. MAP-1 additional captivity/ignorance and overall response asymmetry demonstration. Box-Tukey inverse power transformation – logit family. Standard Box-Cox type (PROBABILITY P-2 AND P-6 algorithms). Choice probability for three modes; business trips (CAA/IPS 1996 data)

I.W.A. ELASTICITIES and (COND. T-STATISTIC)							
	VARIANT =	MAP 1.1	MAP 1.2	D 1.3	D 1.4	D 1.5	D 1.6
	FAMILY =	LOGIT	LOGIT	BT-IPTL	BT-IPTL	BT-IPTL	BT-IPTL
	TYPE =	linlogit	bcllogit	bt_1L1M	bt1L1M1P	bt1L1M3P	bt_1L3M
	PURPOSE =	BUSINESS	BUSINESS	BUSINESS	BUSINESS	BUSINESS	BUSINESS
ALTERNATIVE 1 : AIR							
NETWORK VARIABLES							
TOTAL GROUP COST [AIR]	TXCostA	-.07	-.09	-.09	-.09	-.03	-.08
		(-18.61)	(-20.30)	(-29.56)	(-1.77)	(-2.14)	(-29.69)
	(1)	L 1(GE)	L 1(GE)	L 1(GE)	L 1(GE)	L 1(GE)	L 1(GE)
TOTAL TRAVEL TIME [AIR]	TTTimeA	-.05	-.14	-.15	-.14	-.07	-.14
		(-14.25)	(-23.03)	(-24.91)	(-1.76)	(-2.16)	(-23.75)
	(1)	L 1(GE)	L 1(GE)	L 1(GE)	L 1(GE)	L 1(GE)	L 1(GE)
TOTAL OUT-OF- VEHICLE [AIR]	TOoVTimA	-.01	-.03	-.05	-.06	-.49	-.09
		(-2.35)	(-6.59)	(-9.31)	(-1.74)	(-2.21)	(-15.34)
	(1)	(GE)	(GE)	(GE)	(GE)	(GE)	(GE)
REGRESSION CONSTANT	CONSTANT	-	-	-	-	-	-
		(34.31)	(31.05)	(25.57)	(1.75)	(2.25)	(18.52)
	(1)						
ALTERNATIVE 2 : RAIL							
NETWORK VARIABLES							
TOTAL GROUP COST [RAIL]	TXCostR	-.57	-1.40	-1.34	-1.32	-.66	-1.47
		(-18.61)	(-20.30)	(-29.56)	(-1.77)	(-2.14)	(-29.69)
	(2)	L 1(GE)	L 1(GE)	L 1(GE)	L 1(GE)	L 1(GE)	L 1(GE)
TOTAL TRAVEL TIME [RAIL]	TTTimeR	-1.00	-1.32	-1.30	-1.27	-1.07	-1.58
		(-14.25)	(-23.03)	(-24.91)	(-1.76)	(-2.16)	(-23.75)
	(2)	L 1(GE)	L 1(GE)	L 1(GE)	L 1(GE)	L 1(GE)	L 1(GE)
TOTAL OUT-OF-VEHICLE [RAIL]	TOoVTimR	-.09	-.21	-.30	-.32	-4.03	-.67
		(-2.35)	(-6.59)	(-9.31)	(-1.74)	(-2.21)	(-15.34)
	(2)	(GE)	(GE)	(GE)	(GE)	(GE)	(GE)
TOTAL NUMBER OF TRANSFERS [RAIL]	TUmstR	-1.11	-1.06	-1.26	-1.28	-.32	-.81
		(-14.25)	(-15.93)	(-15.40)	(-1.77)	(-2.08)	(-8.65)
	(2)	(SP)	(SP)	(SP)	(SP)	(SP)	(SP)
REGRESSION CONSTANT	CONSTANT	-	-	-	-	-	-
		(19.09)	(18.43)	(21.24)	(1.76)	(2.25)	(14.99)
	(2)						
ALTERNATIVE 4 : CAR							
NETWORK VARIABLES							
TOTAL COST [CAR]	TCostC	-.76	-1.25	-1.17	-1.22	-1.00	-1.58
		(-18.61)	(-20.30)	(-29.56)	(-1.77)	(-2.14)	(-29.69)
	(4)	L 1(GE)	L 1(GE)	L 1(GE)	L 1(GE)	L 1(GE)	L 1(GE)
TOTAL TRAVEL TIME [CAR]	TTTimeC	-1.12	-1.34	-1.27	-1.27	-1.62	-1.84
		(-14.25)	(-23.03)	(-24.91)	(-1.76)	(-2.16)	(-23.75)
	(4)	L 1(GE)	L 1(GE)	L 1(GE)	L 1(GE)	L 1(GE)	L 1(GE)
TOTAL OUT-OF-VEHICLE [CAR]	TOoVTimC	-0...	-0...	-0...	-0...	-.01	-.01
	--------	(-2.35)	(-6.59)	(-9.31)	(-1.74)	(-2.21)	(-15.34)
	(4)	(GE)	(GE)	(GE)	(GE)	(GE)	(GE)

Table 14.4. (continued)

II. PARAMETERS and UNCOND. [T-STAT.=0] UNCOND. [T-STAT.=1]	VARIANT = FAMILY = TYPE = PURPOSE =	MAP 1.1 LOGIT linlogit BUSINESS	MAP 1.2 LOGIT bcllogit BUSINESS	D 1.3 BT-IPT bt_1L1M BUSINESS	D 1.4 BT-IPT bt1L1M1P BUSINESS	D 1.5 BT-IPT bt1L1M3P BUSINESS	D 1.6 BT-IPT bt_1L3M BUSINESS
BOX-COX TRANSFORMATIONS							
LAMBDA(X) - GROUP 1	LAM 1	1.00 FIXED	-.62 [-13.36] [-34.90]	-.62 [-19.64] [-51.28]	-.49 [-12.58] [-38.38]	-.42 [-6.28] [-21.30]	-.58 [-17.40] [-47.42]
EXTRA PARAMETERS							
PHI 1 AIR				.00 FIXED	0... [3.37] [-1211] GRP 1	0... [3.68] [-880.8]	.00 FIXED
PHI 2 RAIL				.00 FIXED	0... [3.37] [-1211] GRP 1	0... [3.76] [-934.5]	.00 FIXED
PHI 4 CAR				.00 FIXED	0... [3.37] [-1211] GRP 1	0... [3.78] [-963.7]	.00 FIXED
MU 1 AIR				0... [.03] [-INF] GRP 1	0... [.01] [-INF] GRP 1	0... [.01] [-INF] GRP 1	0... [.03] [-INF]
MU 2 RAIL				0... [.03] [-INF] GRP 1	0... [.01] [-INF] GRP 1	0... [.01] [-INF] GRP 1	0... [.03] [-INF]
MU 4 CAR				0... [.03] [-INF] GRP 1	0... [.01] [-INF] GRP 1	0... [.01] [-INF] GRP 1	0... [.03] [-INF]

III. GENERAL STATISTICS	VARIANT = FAMILY = TYPE = PURPOSE =	MAP 1.1 LOGIT linlogit BUSINESS	MAP 1.2 LOGIT bcllogit BUSINESS	D 1.3 BT-IPT bt_1L1M BUSINESS	D 1.4 BT-IPT bt1L1M1P BUSINESS	D 1.5 BT-IPT bt1L1M3P BUSINESS	D 1.6 BT-IPT bt_1L3M BUSINESS
LOG-LIKELIHOOD	-FINAL VALUE	-6731.15	-6246.86	-6233.19	-6216.05	-5481.88	-6025.21
	-WITH CONSTANTS	-7286.77	-7286.77	-7286.77	-7286.77	-7286.77	-7286.77
RHO-SQUARED		.08	.14	.14	.15	.25	.17
RHO-SQUARED BAR	-AKAIKE	.08	.14	.14	.15	.25	.17
	-HOROWITZ	.08	.14	.14	.15	.25	.17
	-HENSHER & JOHNSON	.08	.14	.14	.15	.25	.17
PER CENT RIGHT		92.07	92.19	92.16	92.12	92.87	91.70
SAMPLE	-NUMBER OF ALTERNATIVES	3	3	3	3	3	3
	-NUMBER OF OBS.(ALL MODES)	22579	22579	22579	22579	22579	22579
NUMBER OF ESTIMATED PARAMETERS							
-BETAS	.VARIABLES	4	4	4	4	4	4
	.CONSTANTS	2	2	2	2	2	2
	.ASSOCIATED DUMMIES	0	0	0	0	0	0
-BOX-COX TRANSFORMATIONS		0	1	1	1	1	1
-THETA		0	0	0	0	0	0
-PHI		0	0	0	1	3	0
-MU		0	0	1	1	1	3

Note: under each t-statistic of Section I, the symbols denote generic (GE) or specific (SP) regression coefficient, as well as (L), the presence of a BCT and its group identification.

Table 14.5. MAP-1 additional non-linearity and spatial competition demonstration. BC-DAUHESEQ family (LEVEL 2.0 algorithm). Multiplicative, Box-Cox (8), heteroskedasticity and directed auto-correlation types. Total flow by all modes; Business trips (CAA/IPS 1991 and SES 1993–1994 data)

I. ELASTICITY E(y) (EP) and (COND. T-STATISTIC)	VARIANT FAMILY TYPE PURPOSE	=D 1.1 =LEVEL-2 =LOG =BUSINESS	D 1.2 LEVEL-2 BC8 BUSINESS	D 1.3 LEVEL-2 BC8+ HE BUSINESS	D 1.4 LEVEL-2 BC8+ HE+AU BUSINESS	D 1.5 LEVEL-2 BC8+ HE+AU+PR BUSINESS
ORIGIN CHARACTERISTICS						
WORKING POP. WEIGHTED AT ORIGIN	wpopxpsi3	0.42 (5.62) LAM	0.31 (8.39) LAM	0.30 (10.65) LAM	0.32 (10.05) LAM	0.32 (9.98) LAM
GDP PER WORKER AT ORIGIN	o_gdppwc	0.22 (1.03) LAM	0.38 (1.74) LAM	0.38 (2.60) LAM	0.39 (2.24) LAM	0.39 (2.24) LAM
AREA OF ORIGIN	o_area	0... (0.03) LAM	-0.11 (-1.62) LAM	-0.07 (-1.06) LAM	-0.06 (-0.93) LAM	-0.06 (-0.93) LAM
DESTINATION CHARACTERISTICS						
WORKING POP. AT DESTINATION	d_workpop	0.42 (5.52) LAM	0.42 (7.86) LAM	0.41 (8.13) LAM	0.44 (8.21) LAM	0.44 (8.15) LAM
GDP PER WORKER AT DESTINATION	d_gdppwc	0.42 (1.93) LAM	0.34 (2.12) LAM	0.33 (2.28) LAM	0.29 (1.88) LAM	0.29 (1.85) LAM
AREA OF DESTINATION	d_area	-0.04 (-0.60) LAM	-0.08 (-1.29) LAM	-0.07 (-1.32) LAM	-0.08 (-1.52) LAM	-0.08 (-1.50) LAM
IMPEDANCES						
UTILITY	util_int3	0.11 (4.36) LAM	0.16 (5.95) LAM	0.15 (5.66) LAM	0.16 (5.62) LAM	0.16 (5.60) LAM
BORDER IS CROSSED	cross =====	-0.63 (-1.94)	-0.61 (-1.88)	-0.50 (-1.52)	-0.55 (-1.63)	-0.55 (-1.63)
DATABASE						
CHANNEL SURVEY	ips_caa =======	-2.49 (-7.96)	-1.92 (-5.67)	-2.04 (-6.18)	-1.95 (-5.75)	-1.95 (-5.75)
REGRESSION CONSTANT	CONSTANT	- (0.98)	- (5.48)	- (6.07)	- (4.84)	- (4.76)
HETEROSKEDASTICITY STRUCTURE						
DELTA COEFFICIENTS						
DISTANCE BY CAR	dist			-0.26 (-5.03) LAM	-0.23 (-4.40) LAM	-0.23 (-4.39) LAM

Table 14.5. (continued)

II. PARAMETERS and (COND. T-STATISTIC)		VARIANT =D 1.1	D 1.2	D 1.3	D 1.4	D 1.5
		FAMILY =LEVEL-2	LEVEL-2	LEVEL-2	LEVEL-2	LEVEL-2
		TYPE =LOG	BC8	BC8+ HE	BC8+ HE+AU	BC8+ HE+AU+PR
		PURPOSE =BUSINESS	BUSINESS	BUSINESS	BUSINESS	BUSINESS
HETEROSKEDASTICITY STRUCTURE						
BOX-COX TRANSFORMATIONS: UNCOND: [T-STATISTIC=0] / [T-STATISTIC=1]						
LAMBDA(Z)	dist			0.86 [1.25] [-0.20]	0.90 [1.17] [-0.14]	0.90 [1.17] [-0.14]
BOX-COX TRANSFORMATIONS: UNCOND: [T-STATISTIC=0] / [T-STATISTIC=1]						
LAMBDA(Y)	flow	0.00 FIXED	0.03 [1.56] [-47.77]	0.07 [2.53] [-35.13]	0.06 [2.33] [-35.56]	0.06 [2.28] [-34.90]
LAMBDA(X)	wpopxpsi3	0.00 FIXED	1.59 [3.32] [1.23]	1.76 [3.92] [1.70]	1.62 [3.87] [1.48]	1.62 [3.86] [1.48]
LAMBDA(X)	o_gdppwc	0.00 FIXED	-1.15 [-0.54] [-0.90]	-0.43 [-0.28] [-0.93]	-0.59 [-0.34] [-0.90]	-0.60 [-0.34] [-0.90]
LAMBDA(X)	o_area	0.00 FIXED	0.75 [0.62] [-0.21]	0.85 [0.44] [-0.08]	0.92 [0.42] [-0.03]	0.92 [0.41] [-0.03]
LAMBDA(X)	d_workpop	0.00 FIXED	0.42 [1.74] [-2.37]	0.45 [1.80] [-2.20]	0.51 [2.06] [-2.00]	0.51 [2.06] [-2.00]
LAMBDA(X)	d_gdppwc	0.00 FIXED	0.78 [0.51] [-0.15]	0.80 [0.56] [-0.14]	0.87 [0.49] [-0.07]	0.87 [0.49] [-0.07]
LAMBDA(X)	d_area	0.00 FIXED	0.09 [0.10] [-0.95]	-0.07 [-0.08] [-1.21]	-0.07 [-0.09] [-1.35]	-0.07 [-0.09] [-1.35]
LAMBDA(X)	util_int3	0.00 FIXED	0.25 [3.67] [-11.08]	0.25 [3.76] [-11.11]	0.25 [3.67] [-11.24]	0.25 [3.66] [-11.23]
SPATIAL AUTOCORR.: COND: [T-STATISTIC=0] / [T-STATISTIC=1]				O/D Dist (100-300km)		
RHO (dist)					0.18 (2.56)	0.18 (1.97)
PI (dist)					1.00 FIXED	0.98 [1.91] [-0.04]

III. GENERAL STATISTICS	VARIANT =D 1.1	D 1.2	D 1.3	D 1.4	D 1.5
LOG-LIKELIHOOD	-4890.33	-4868.30	-4858.08	-4852.16	-4852.16
SAMPLE: NUMBER OF PAIRS	544	544	544	544	544
NUMBER OF ESTIMATED PARAMETERS :					
- FIXED PART :					
. BETAS	10	10	10	10	10
. BOX-COX	0	8	8	8	8
. ASSOCIATED DUMMIES	0	0	0	0	0
- AUTOCORRELATION					
. RHO	0	0	0	1	1
. PI	0	0	0	0	1
- HETEROSKEDASTICITY :					
. DELTAS	0	0	1	1	1
. BOX-COX	0	0	1	1	1
. ASSOCIATED DUMMIES	0	0	0	0	0

is present. If one removes the equality constraint imposed on the individual $\widetilde{\mu}_i$, further gains of 7 points are achieved (comparing column D 1.3 and column D 1.6). These gains suggest that the modeler's ignorance differs across the modes.

Overall asymmetry of response. Starting from column D 1.3 where captivity/ ignorance is established, the addition of a common $\widetilde{\varphi}$ on all three alternatives in column D 1.4 causes the log-likelihood to increase by 17 points: this large increase demonstrates the presence of overall response asymmetry even if the optimal value of the $\widetilde{\varphi}$ – indicated in Table 14.4 as { 0...} but in fact more precisely stated as {0,003} – is close to zero (with a *t*-statistic of only 3.37 for the numerical reasons mentioned above). This means that, despite accounting for non-linearity on the X_n, there remains non-linearity in the response to the V_i function itself, as well as modeler's ignorance in the MAP-1 business model.

Demonstration of New Dimensions: Generation-Distribution. In Table 14.5, the first column D 1.1 shows estimates obtained with the same specification as that found in column MAP 1.1 of Table 14.3, but with a sample reduced randomly from 5711 to 544 observations in a way that preserves the shares of the 5 data sources. Although reduction in sample size would not be necessary to search for additional non-linearity, it is practical in a search for spatial autocorrelation of the residuals because of the need to compute repeatedly throughout estimation the inverse of matrix $\widetilde{R}_l$ defined in (8), of size T by T determined by the number of observations. We then start again with the classical multiplicative gravity equation results of column D 1.1 that differ slightly from those of the ancestor model MAP 1.1 (in particular, the 'weighted population at origin' variable is now strictly positive, removing the need for a compensating associated dummy variable).

Generalized gravity with heteroskedasticity. In contrast with the case shown in Table 14.3, where three BC are used, we now use a different BC on all transformable variables (true dummy variables cannot be transformed), dependent and independent, extracting further non-linearity than if we had used only 3 BC (the test value is not shown). We then add in column D 1.3 the generalized heteroskedasticity form with the distance variable previously used. The elasticity of the distance variable implies that the variance of the error term falls with distance, but at a decreasing rate (the BC of distance equals 0.86).

Spatial autocorrelation. In column D 1.4, we test for spatial correlation of residuals under the rule that the typical element r_{tn} of the R matrix for a first order process be equal to one if either the flow considered has the same destination and is a near neighbour of the origin, or conversely (has the same origin and is a near neighbour of the destination), and otherwise be equal to zero. This rule, applied with a definition of 'near' to mean 'any point in *a ring situated from 100 to 300 km* from the point considered', states that one expects errors of explanation of flows with starting or destinating points 'near' (within the rings) those considered to be correlated.

We did not choose circles around each point, but rings, in order to exclude any remaining short distance trips from consideration. We chose 100-300 km, instead of the 100-180 km used for Germany in the baseline model because of the multicountry nature of the application. Results of column D 1.4 show that positive spatial correlation, indicative of competition among markets, is present, but only for close neighbours, as the proximity parameter π that defines the tail remains almost equal to 1 in column D 1.5.

Both non-linearity and spatial correlation are then still present in the MAP-1 business model.

14.4 Conclusion

We have described the first-ever multicountry application for passengers of the QDF format, obtained using the TRIO baseline algorithms: MAP-1 consists in a combination of disaggregate mode choice models and aggregate total demand models pertaining to three trip purposes. We also performed advanced tests of the algorithmic improvements effected under STEMM and demonstrated their usefulness on the business trip components of the MAP-1 multicountry application within the QDF structure.

Acknowledgements

This paper is based on research financed predominantly by the 'STrategic European Multimodal Model (STEMM) Project' (as part of Working Package 4A: Software design and implementation - Passengers), Contract No.: ST-96-SC.301 of the Commission of the European Communities, with background support provided by Marc Gaudry's tenure as 1995-1997 Ernst Blickle Research Prize awardee at Universität Karlsruhe and as 1998 Centre National de la Recherche Scientifique (CNRS) researcher at BETA, Université Louis Pasteur and UMR CNRS 7522.

References

Berndt, E.R., Hall, B.H., Hall, R.E. and Hausman, J.A. (1974), 'Estimation and inference in nonlinear models', *Annals of Economic and Social Measurement*, vol. 3, pp. 653-665.

Box, G.P. and Cox, D.R. (1964), 'An analysis of transformations', *Journal of the Royal Statistical Society*, Series B, vol. 26, pp. 211-243.

COST 305 (1988), 'Data system for the study of demand for interregional passenger transport', EUR 11443 EN, Commission of the European Communities, Brussels.

Eberhard, C., Heinitz, F., Last, J., Mandel. B., Rothengatter, W. and Schoch, M. (1998), 'An integrated framework for the STEMM passenger model', Annex A, Final Consolidated Progress Report, STEMM, Contract No.: ST-96-SC.301 of the Commission of the European Communities, Brussels.

EUROSTAT (1992), 'Regions: Nomenclature of territorial units for statistics—NUTS', Statistical Office of the European Communities, Luxemburg.

Gaudry, M. (1981), 'The inverse power transformation logit and dogit mode choice models', *Transportation Research B*, vol. 15, pp. 97-103.

Gaudry, M. (with many colleagues) (1997), 'Cur cum TRIO?', Publication CRT-901, Centre de recherche sur les transports, Université de Montréal, Montreal.

Gaudry, M., Heinitz, F., Last, J. and Mandel, B. (1998), 'Methodological developments within the quasi-direct format demand structure: The multicountry application for passengers MAP-1',

Publication CRT-98-46, Centre de recherche sur les transports, Université de Montréal, Montral, and Working Paper N° 9815, Bureau d'économie théorique et appliquée (BETA), Université Louis Pasteur, Strasbourg.

Gaudry, M., Mandel, B. and Rothengatter, W. (1994), 'Introducing spatial competition through an autoregressive contiguous distributed (AR-C-D) process in intercity generation-distribution models within a quasi-direct format (QDF)', Publication CRT-971, Centre de recherche sur les transports, Université de Montréal, Montreal.

Gaudry, M. and Wills, M.J. (1978), 'Estimating the functional form of travel demand models', *Transportation Research*, vol. 12, pp. 257-289.

INSEE (1993), 'Various questionnaires: "Enquête Transports 1993-1994", or "Une enquête par sondage sur les transports en 1993-1994" ', Institut National de la Statistique et des Etudes Économiques, Paris.

Last, J. (1998), 'Reisezweckspezifische Modellierung von Verkehrsverflechtungen', Institut für Wirtschaftspolitik und Wirtschaftsforschung (IWW), and MKmetric GmbH, Karlsruhe.

McLynn, J.M., Goldman, A.J., Meyers, P.R. and Watkins, R.H. (1968), 'Analysis of a market split model', *Journal of Research of the National Bureau of Standards*, vol. 72-B, pp. 32-51.

McLynn, J.M., and Woronka, T. (1969), *A Family of Demand and Mode Split Models*, Arthur Young and Co., New Haven.

List of Figures

List of Tables

Subject Index

Subject Index

Contributors

Staffan Algers	Transek AB, Solna Torg 3, SE-171 45 Solna, Sweden
Muriel Beser	Transek AB, Solna Torg 3, SE-171 45 Solna, Sweden
Isabelle Constantin	INRO Consultants Inc., 5160 Décarie Boulevard, Suite 610, Montréal, Canada H3X 2H9
Andrew Daly	RAND Europe, Newtonweg 1, 2333 CP Leiden, The Netherlands
Michael Florian	INRO Consultants Inc., 5160 Décarie Boulevard, Suite 610, Montréal, Canada H3X 2H9
Mogens Fosgerau	Danish Transport Research Institute, Knuth-Winterfeldts Allé, Bygning 116 Vest DK-2800 Kgs. Lyngby, Denmark
Marc Gaudry	Bureau d'économie théorique et appliquée (BETA), Université Louis Pasteur, Strasbourg, France and Centre de recherche sur les transports (CRT) Université de Montréal, Montréal, Canada
Hugh Gunn	RAND Europe, Newtonweg 1, 2333 CP Leiden, The Netherlands
Russell C.E. Harris	Highways Economics and Traffic Appraisal Division Department of the Environment, Transport and the Regions Great Minster House, 76 Marsham Street, London SW1P 4DR, United Kingdom

Frank Hofman	Ministry of Transport, Public Works and Water Management Directorate-General of Public Works and Water Management Transport Research Centre (AVV), Strategic Studies Division Rotterdam, The Netherlands
Gunnar Johansson Sveder	Transek AB, Solna Torg 3, SE-171 45 Solna, Sweden
Lars Lundqvist	Department of Infrastructure and Planning Division of Transport and Location Analysis Royal Institute of Technology, SE-100 44 Stockholm, Sweden
Lars-Göran Mattsson	Department of Infrastructure and Planning Division of Transport and Location Analysis Royal Institute of Technology, SE-100 44 Stockholm, Sweden
Francesco Russo	Department of Science, Mathematics, Electronics and Transportation University of Reggio Calabria, 89100 Reggio Calabria, Italy
Heinz Spiess	INRO Consultants Inc., 5160 Décarie Boulevard, Suite 610 Montréal, Canada H3X 2H9
Staffan Widlert	Swedish Institute for Transport and Communications Analysis, SIKA, P.O. Box 17 213, SE-104 62 Stockholm, Sweden
Ian Williams	Marcial Echenique & Partners Ltd, 49-51 High Street, Trumpington, Cambridge CB2 2HZ, United Kingdom
Tom E. Worsley	Highways Economics and Traffic Appraisal Division Department of the Environment, Transport and the Regions Great Minster House, 76 Marsham Street, London SW1P 4DR, United Kingdom

GPSR Compliance

The European Union's (EU) General Product Safety Regulation (GPSR) is a set of rules that requires consumer products to be safe and our obligations to ensure this.

If you have any concerns about our products, you can contact us on ProductSafety@springernature.com

In case Publisher is established outside the EU, the EU authorized representative is:

Springer Nature Customer Service Center GmbH
Europaplatz 3
69115 Heidelberg, Germany

Batch number: 10755292

Printed by Printforce, the Netherlands